SpringerBriefs in Geotechnical and Earthquake Engineering

Series editor

Atilla Ansal, Özyeğin University, Istanbul, Turkey

The Aims of This New Series

Springer Briefs in Geotechnical and Earthquake Engineering is to establish an authoritative set of publications covering the most recent findings and design procedures in these engineering fields. Volumes are presented in an easy accessible and compact form to researchers and practitioners to advance the state-of-the-art and state-of-the-practice. The volumes will be peer-reviewed and be overviewed by a respected editorial board.

Typical topics might include: *Topical overviews, case studies and applications in the fields of soil mechanics, foundation engineering, geotechnical earthquake engineering, earthquake engineering, earthquake hazard and other related subjects.*

More information about this series at http://www.springer.com/series/15951

Florin Pavel · Viorel Popa
Radu Vacareanu

Impact of Long-Period Ground Motions on Structural Design: A Case Study for Bucharest, Romania

 Springer

Florin Pavel
Seismic Risk Assessment Research Center
Technical University of Civil Engineering
 Bucharest (UTCB)
Bucharest
Romania

Radu Vacareanu
Seismic Risk Assessment Research Center
Technical University of Civil Engineering
 Bucharest (UTCB)
Bucharest
Romania

Viorel Popa
Seismic Risk Assessment Research Center
Technical University of Civil Engineering
 Bucharest (UTCB)
Bucharest
Romania

ISSN 2522-8781   ISSN 2522-8803   (electronic)
SpringerBriefs in Geotechnical and Earthquake Engineering
ISBN 978-3-319-73401-9   ISBN 978-3-319-73402-6   (eBook)
https://doi.org/10.1007/978-3-319-73402-6

Library of Congress Control Number: 2017963291

Printed on acid-free paper

This Springer imprint is published by Springer Nature
The registered company is Springer International Publishing AG
The registered company address is: Gewerbestrasse 11, 6330 Cham, Switzerland

Preface

The issue of long-period strong ground motions arose for the first time in the second half of the twentieth century in Japan. Since then, compelling evidence of long-period seismic waves has occurred in many strong ground motion recordings obtained throughout the world. The only strong ground motion recorded in Bucharest during the 4 March 1977 Vrancea intermediate-depth earthquake revealed, for the first time in Romania, low-frequency (long-period) spectral components of ground motion, which were responsible for the heavy damage or complete collapse of tall, flexible and vulnerable residential buildings in Bucharest. Since then, the earthquake of 30 August 1986, which was also generated by the Vrancea intermediate-depth seismic source, produced strong ground motions with low-frequency content in some sites in Bucharest. It should be noted that the Michoacan earthquake of 19 September 1985 produced in Mexico City the narrowest and lowest frequency content strong ground motion recorded until then. Later on, Munich Re. who compiled "The World Map of Natural Hazards" named Bucharest a "large city with Mexico-City effects".

The issue of long-period ground motions was from the beginning at the focus of engineering seismology, and later came to the attention of earthquake engineering specialists. The challenges raised to structural engineering by the very high seismic displacement demands are daunting, and the technical solutions to tackle these challenges are, sometimes, very costly. Nevertheless, it is at the very core of the structural engineering to find solutions to such complicated problems.

In Bucharest, the compelling evidences of the issue of long-period ground motions are provided by actual records. As a consequence, the design acceleration response spectrum has a very long control period T_C of 1.6 s (among the largest values in the world). The design peak ground acceleration with 20% exceedance probability in 50 years is 0.3 g, and it is envisaged that this will be raised by another 20–25% in the next revision of the Romanian earthquake-resistant design code (to codify design values with 10% exceedance probability in 50 years). The issue of the control period T_C increasing along with magnitude increasing is a hot topic, as well. Consequently, the seismic demands in terms of displacements, that are quite high at present, will further increase, thus exacerbating the challenges encountered by the structural engineers. Nevertheless, the need for a larger and safer building stock in Bucharest must be properly addressed both by designers and contractors, as well. The code drafters cannot argue with nature, so the issue of long-period high-spectral acceleration design spectra is clearly stated in the documents issued and enforced by the building officials.

To the best of the authors' knowledge, the combination of long-period and high-acceleration design spectrum is not common in seismic codes and regulations, so there is limited experimental experience or practical expertise in the design of flexible or high-rise buildings for this kind of challenging seismic demand. Up to now, the Romanian design practice has found some solutions to meet the concurrent demands for very high strength, deformability and ductility. Nevertheless, it seems that the boundaries of structural earthquake engineering are pushed by the current solutions: very high density of shear walls, very heavy RC cores, very heavily reinforced structural elements and very thick foundation mats. Of course, it is legitimate to ask about the employment of seismic isolation devices and energy dissipation solutions. The former is limited by the very high displacement demands. Certainly, dampers and friction pendulums might reduce these displacement demands (albeit, the strength demand is increased), but the cost is an important criterion on a very competitive real-estate market, as it is the case in Bucharest. Moreover, when applying base isolation, one has to consider that the horizontal-to-vertical spectral ratio method applied on ambient vibrations in Bucharest revealed spectral peaks for periods of five to six seconds. There is limited opportunity for the employment of energy dissipation devices because of the associated cost and limitation of functionality and usage class of the buildings.

The recent experience of the 14 November 2016 Kaikoura earthquake in New Zealand raised again the problem of seismic damage produced by long-period strong ground motions in Wellington. One has to put this problem in the context of the paradigm shift in seismic design towards resilience. This latter approach is further complicating the earthquake-resistant design of structural and non-structural elements for long-period ground motions. It is the belief of the authors that the international community of seismologists and engineers will tackle this issue properly and the need for development will be fuelled by their collective wisdom. The authors consider this publication as a golden opportunity to share the Romanian experience of long-period high-acceleration strong ground motions, both in engineering seismology and earthquake engineering.

This book in the Springer Brief series aims at discussing the impact of long-period ground motions on structural design using as case study the situation of Bucharest, the capital city of Romania. The first part of the Brief evaluates the soil conditions in the Bucharest area. The characteristics of ground motions for Bucharest in terms of amplitudes and frequency contents of both recorded and simulated motions are subsequently discussed. The causes of long-period ground motions related to both source and site features are then discussed in the light of a new ground motion model specifically derived for Bucharest using a ground motion database of both natural and simulated ground motion recordings. Next, a discussion regarding the current seismic design and detailing practice for buildings built in Bucharest is presented. The evolution of seismic design codes for Bucharest and the assessment of inelastic spectral displacements are also addressed in the same section. Moreover, the impact of long-period ground motions on the seismic design practice is analysed. Finally, several case studies of buildings in Bucharest are presented, and the major challenges encountered in their design are discussed. This Brief contains various numerical examples which will aid the reader to better understand the discussed topics. The book is addressed to both researchers in the field of seismic hazard and risk assessment and designers of multi-storey buildings located in moderate and high seismic areas.

In the opinion of the authors, out of the numerous national and international projects aiming at understanding the Romanian seismicity and at improving the codes and regulations for seismic design and evaluation of buildings, the most comprehensive one was the Japan International Cooperation Agency (JICA) Technical Cooperation Project for Seismic Risk Reduction, implemented in the period 2002–2008. Many outstanding scholars, researchers and engineers from Japan served as short-term and long-term experts in Romania during that period and became acquainted with the issues tackled in this book. Two of the experts, Dr. Toshihide Kashima, senior researcher at the International Institute for Seismology and Earthquake Engineering of Building Research Institute, Tsukuba, and Dr. Koichi Kusunoki, associate professor at the Earthquake Research Institute, University of Tokyo, gracefully accepted to review the manuscript of this Brief. The authors would like to extend their gratitude to the reviewers; their valuable comments and suggestions are very much appreciated as they have helped us to greatly improve the quality of the manuscript.

Some of the results shown in this Brief were obtained within the framework of three national research projects implemented at the Seismic Risk Assessment Research Center of Technical University of Civil Engineering of Bucharest in the period 2012–2017, namely BIGSEES, COBPEE and RO-RISK. The authors extend their gratitude to the Romanian taxpayers for financing these grants.

The authors would also like to thank Petra van Steenbergen and all the editorial staff from Springer International Publishing AG for their professional coordination and support in preparing this work in Springer Briefs in Geotechnical and Earthquake Engineering.

Bucharest, Romania   Florin Pavel
August 2017   Viorel Popa
                                                                                       Radu Vacareanu

Contents

Chapter 1
Introduction

Long-period ground motions and their effects on the built environment have come to the attention of the scientific community especially after the devastating Michoacan (Mexico) earthquake of September 1985. The losses caused by this earthquake, both in terms of structural damage and human casualties, were extensive in an area of Mexico City that overlays an old lake bed. However, long-period ground motions had already been identified in other parts of the world. For instance, Koketsu and Miyake (2008) note that long-period ground motions were identified for the first time during the 1968 Tokachi-Oki earthquake at Hachinohe station in Japan (the predominant period of soil vibration was around 2.5 s). The authors associate these types of far-field long-period ground motions to offshore earthquakes occurring in subduction zones. However, inland earthquakes can also generate long-period ground motions due to a combination of factors, such as: rupture directivity, basin effects and/or site effects can be responsible for this type of ground motions.

Romania, and specifically its southern part (including Bucharest), has been affected by numerous large-magnitude intermediate-depth earthquakes originating in the Vrancea intermediate-depth seismic source. The only free-field ground motion that was recorded in Romania during the March 4, 1977 Vrancea earthquake at a seismic station in the eastern part of Bucharest showed very large spectral amplifications for periods of vibration in the range 1.0–2.0 s. However, in the case of this ground motion recording, besides the site effects, another element that increased the level of damage, especially in the Bucharest area, was the velocity of the earthquake rupture propagation that was almost equal to that of the shear waves (Hartzell 1979).

In April 2015, the strong ground motions recorded during the Gorkha earthquake (Nepal) have shown that through a combination of basin effects and directivity, the predominant period of the acceleration response spectra can be in excess of 5 s for both horizontal components, leading to very large displacement demands for high-rise/flexible buildings/structures.

© The Author(s) 2018
F. Pavel et al., *Impact of Long-Period Ground Motions on Structural Design:
A Case Study for Bucharest, Romania*, SpringerBriefs in Geotechnical
and Earthquake Engineering, https://doi.org/10.1007/978-3-319-73402-6_1

Consequently, due to the combination of source and site effects, large spectral displacement demands are to be expected during large-magnitude Vrancea earthquakes. In this book we will provide some estimates regarding these large displacement demands and we will put forward the corresponding issues raised in earthquake resistant design practice.

References

Hartzell S (1979) Analysis of the Bucharest strong ground motion for the March 4, 1977 Romanian earthquake. B Seismol Soc Am 69:513–530
Koketsu K, Miyake H (2008) A seismological overview of long-period ground motion. J Seismol 12:133–143

Chapter 2
Evaluation of Soil Conditions in Bucharest

Abstract In this chapter, the soil conditions in the Bucharest area are evaluated using several approaches based on both observed and simulated ground motion recordings. The analysis of the recorded ground motions shows the fact that significant long-period spectral ordinates are encountered for large magnitude Vrancea intermediate-depth seismic events. A soil class C according to EN 1998-1/2004 can be assigned for Bucharest based on the shear-wave velocities of the existing boreholes in this area. However, the much deeper Quaternary sediments from the Bucharest area can generate significant long-period spectral ordinates largely exceeding the ones for a typical soil class C site. The nonlinear ground response analysis performed for INCERC site in the eastern part of Bucharest revealed ground motions with a frequency content similar to the one of the ground motion recorded during the Vrancea 1977 earthquake. Besides the peak in amplitude corresponding to a spectral period of around 1.5 s, some additional analyses have shown that another peak at around 5 s can be inferred from the available data.

Keywords Vrancea intermediate-depth seismic source · Shear-wave velocity
Soil layers · Spectral accelerations · Ground response analysis · Amplification
factors · Horizontal-to-vertical spectral ratio

Bucharest, the capital city of Romania, has a population of roughly two million inhabitants that live in an area of about 240 km^2. The city is divided into six Districts that have different building stock characteristics, as revealed by the most recent census conducted in 2011. Two of the Districts, namely 1 and 5, consist mainly of low-rise masonry or adobe buildings, while Districts 3 and 6 have the largest percentage of high-rise RC buildings that were built during the communist period and now shelter most of the residents of Bucharest.

Bucharest has been historically affected by large-magnitude earthquakes originating in the Vrancea intermediate-depth seismic source that is situated at the bend of the Carpathian Mountains. The range of epicentral distances between Bucharest and the likely epicenter locations within the Vrancea intermediate-depth seismic source is between 80 and 180 km. The focal depths of Vrancea intermediate-depth

© The Author(s) 2018
F. Pavel et al., *Impact of Long-Period Ground Motions on Structural Design:
A Case Study for Bucharest, Romania*, SpringerBriefs in Geotechnical
and Earthquake Engineering, https://doi.org/10.1007/978-3-319-73402-6_2

seismic events are between 60 and 220 km (albeit there is a net decrease in seismic activity for focal depths in excess of 160 km). The largest number of earthquakes occur in the focal depth ranges of 90–110 km and 130–150 km. The historic earthquakes of 1738 (estimated moment magnitude M_W = 7.7 and estimated focal depth h = 130 km), 1802 (estimated M_W = 7.9 and estimated h = 150 km) and 1838 (estimated M_W = 7.5 and estimated h = 150 km) caused both damage to buildings and casualties among the Bucharest residents. In the XXth century, the two largest earthquakes occurred in November 1940 (M_W = 7.7 and h = 150 km) and March 1977 (M_W = 7.4 and h = 94 km). The first earthquake, that represents the largest intermediate-depth seismic event occurring in Europe in the XXth century, caused an unknown number of casualties (between 500 and 1000, according to various sources) and the collapse of the tallest reinforced concrete (RC) building at that time in Romania, namely the Carlton Building situated in downtown Bucharest. The second seismic event, albeit of smaller magnitude, but occurring at a shallower depth and at a smaller epicentral distance as compared to the 1940 event, generated much more damage (32 medium- and high-rise buildings with reinforced concrete or masonry structures collapsed in Bucharest) and many more casualties (over 1500 deaths, the majority of which were in Bucharest) than the previous large magnitude event. Many of the high-rise buildings that collapsed during the 1977 earthquake had been affected considerably during the previous large Vrancea seismic event of 1940. The only free-field ground motion recorded during the March 1977 Vrancea earthquake at INCERC station, in the eastern part of Bucharest, revealed significant long-period spectral ordinates that exceeded the design values in force at that moment several times. Significant long-period spectral ordinates were also observed during the subsequent Vrancea earthquake of August 1986 (M_W = 7.1 and h = 131 km). The Vrancea seismic events of May 1990 (two individual events with M_W = 6.9 and h = 91 km and M_W = 6.4 and h = 87 km, that took place in an interval of roughly 12 h) and October 2004 (M_W = 6.0 and h = 105 km) did not reveal the same pattern of significant long-period spectral amplitudes as noted above. The World Map of Natural Hazards compiled by Münich Re (1998) indicates that Bucharest can be considered as a "large city with Mexico-city effect".

Figure 2.1 shows the mean normalized acceleration response spectra (ratio of the spectral accelerations to the corresponding peak ground acceleration for a damping ratio of 5%) for the ground motions recorded in the Bucharest area during five earthquakes with $M_W \geq$ 6.0 that occurred in the period 1977–2004. The number of ground motions recorded during each of the five seismic events varies between one (1977 event) and 20 (2004 event). One can observe that the predominant period of the response spectra decreases with the earthquake magnitude and the spectral peaks occurring at periods of 1.2–1.5 s are not visible for the earthquakes with $M_W \leq$ 6.9. The largest peak ground accelerations in the Bucharest area were recorded during the Vrancea seismic event of May 1990 (M_W = 6.9 and h = 91 km), namely 0.24 g at Peris station and 0.22 g at Bolintin-Vale station. A similar value of 0.22 g was also recorded at Otopeni station in the northern part

Fig. 2.1 Mean normalized acceleration response spectra for ground motions recorded in the Bucharest area during intermediate-depth Vrancea earthquakes

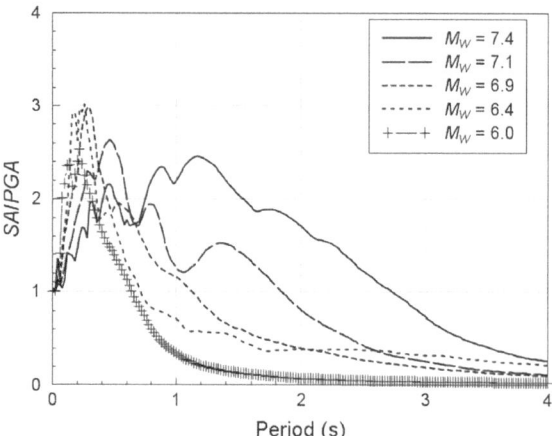

of Bucharest during the Vrancea earthquake of August 1986. In terms of spectral acceleration, the second largest (-1.2 g at a spectral period of roughly 0.3 s) from the entire database of ground motion recordings in Romania was observed during the May 30, 1990 earthquake at Peris station outside Bucharest.

In terms of its geography, Bucharest is situated in the Romanian Plain and has a slight inclination in the north-south direction. Bala et al. (2011) note two special features of the Bucharest geology, namely the bedrock situated at depths in excess of 500–700 m and the alternation down to 300 m in depth of Quaternary sand and clay layers that also contain three aquifer systems. The lithology of the Quaternary soil layers in the Bucharest area defined by Lungu et al. (1999) and Bala et al. (2011) is as follows:

- Layer 1—Backfill and organic soil with depths in the range 0.3–3 m (although it may reach up to 10 m in depth);
- Layer 2—Sandy clay superior deposits—deposits of loess, sandy clay and sands with depths in the range 2–16 m (towards the southern part of the city);
- Layer 3—Colentina gravel complex that also contains an aquifer is a layer consisting of gravels and sands with various grain size distribution. Its thickness is in the range 2–20 m according to Lungu et al. (1999). Bala et al. (2011) mentions that the layer is completely missing in the western part of the city;
- Layer 4—Intermediate clay layer—consists of predominantly clay layers with various consolidation degrees and thickness of up to 25 m;
- Layer 5—Mostistea sandbank—comprises medium and fine sand layers and spreads throughout the whole city. Its thickness can reach 25 m and it contains the second aquifer system;
- Layer 6—Lacustrine complex—layers of marled clay and fine sands with thicknesses in the range 60–130 m, according to Bala et al. (2011). The variable thickness, that increases from south to north, is attributed to the underlying Fratesti complex of gravels;

- Layer 7—Fratesti gravel complex consists of three layers of gravel and sands (10–40 m thick) separated by layers of clay. Its total thickness is in the range 100–180 m and it also contains the third aquifer system. This layer, as previously mentioned, dips towards the north and can be found at depths of around 80 m in the southern part of the city and at depths in excess of 180 m in the northern part of Bucharest.

Bala et al. (2011) compare the mean shear-wave velocities for the seven Quaternary layers in Bucharest obtained using various methods. The results show values that start from around 70 m/s for the backfill layer and reach 500 m/s for the Fratesti gravel complex. Because of these low shear-wave velocities, the Fratesti complex cannot be considered as the bedrock layer of Bucharest. Figures 2.2 and 2.3 show the distribution of the average shear wave velocities on the first 30 m of soil deposits ($v_{s,30}$) and on the first 50 m of soil deposits ($v_{s,50}$) respectively, using borehole data collected within the BIGSEES research project. The range of values for $v_{s,30}$ is mainly in the range of soil class C as defined in EN 1998-1/2004, with the larger values corresponding to soil class B ($v_{s,30}$ values larger than 360 m/s). One can also notice the small differences in the range of values for $v_{s,30}$ and $v_{s,50}$, respectively, thus denoting a small increase of the shear wave velocities with the depth. This observation is subsequently confirmed using data from the deeper boreholes within the Bucharest area.

To enable the reader to get a better grasp of the shear-wave velocity profiles in the Bucharest area, two examples are shown in Fig. 2.4. The depths of the two analysed boreholes (INCERC denoted as INC and VIC near Piata Victoriei) are 153 and 205 m, respectively. The position of the two boreholes is shown in Fig. 2.3. The average shear-wave velocities for the first 30 m of soil deposits ($V_{s,30}$) shows that soil category C (defined as in EN 1998-1/2004) can be assigned to both sites. As previously mentioned, the shear-wave velocities are below 500 m/s even at depths in excess of 150 m. In addition, no clear velocity contrast is apparent from the available data. Consequently, the depth of the bedrock for the Bucharest area can be considered in excess of 200 m.

The amplification ratios computed using the ground motions recorded at various depths in boreholes is another method for evaluating the influence of various soil deposits on the frequency content characteristics. These ratios are computed using the ground motions recorded at INCERC station, in the eastern part of Bucharest, during four moderate magnitude intermediate-depth Vrancea earthquakes, namely the seismic events of October 27, 2004 ($M_W = 6.0$, $h = 105$ km), May 14, 2005 ($M_W = 5.5$, $h = 149$ km), June 18, 2005 ($M_W = 5.2$, $h = 154$ km) and April 25, 2009 ($M_W = 5.4$, $h = 110$ km). The resulting amplification ratios are shown in Figs. 2.5, 2.6, 2.7 and 2.8, using the two recorded horizontal components at surface level, at a depth of 24 m and at a depth of 153 m, which represents the bottom of the borehole. It is interesting to note that the upper part of the borehole (from surface level to a depth of 24 m) is responsible for the amplifications of higher frequencies, while the frequencies below 1 Hz are amplified by the deeper soil deposits. The amplification peak at a period around 0.7 Hz, which is visible on all

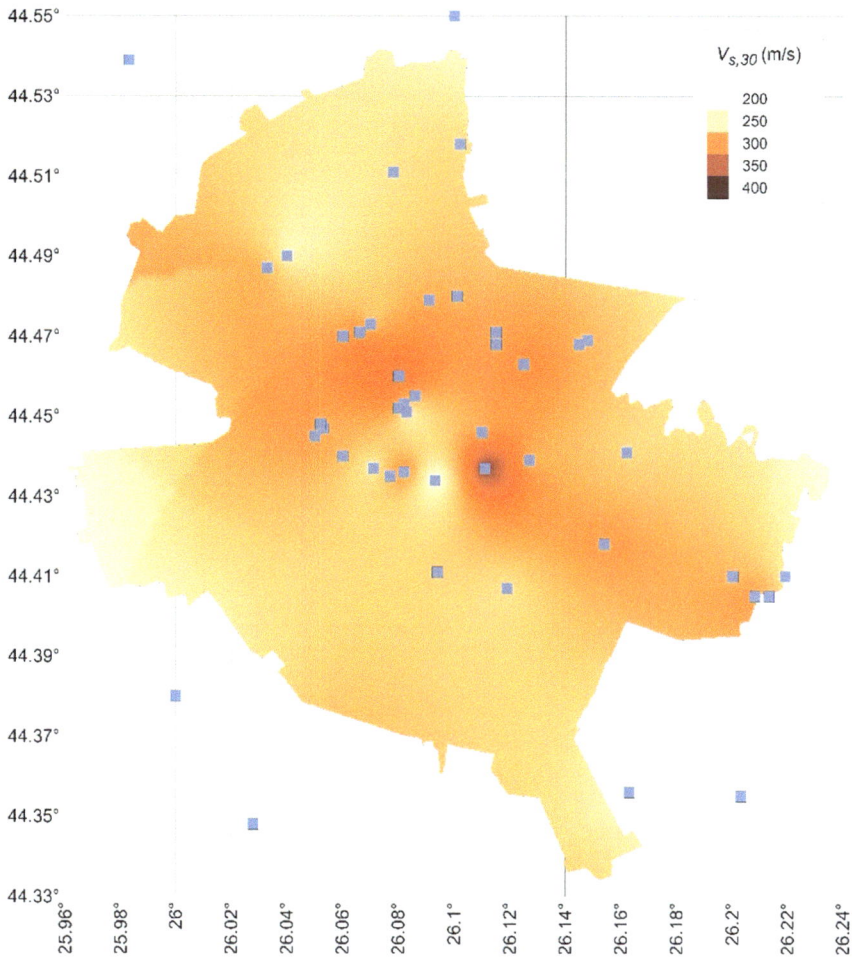

Fig. 2.2 Average shear wave velocities on the first 30 m of soil layers for the available boreholes in the Bucharest area (results from BIGSEES project)

the computed ratios (ratios involving the deeper part of the borehole), is another noteworthy aspect of the analyses.

Several researchers (Grecu et al. 2007; Zaharia et al. 2008; Manea et al. 2016; Pavel 2017) have analysed the horizontal-to-vertical spectral ratios (*HVSR*) for various sites within the Bucharest area. The *HVSR* method, originally proposed by Nakamura (1989), was applied to both ambient vibrations and earthquake recordings. Generally, the results coincide in the sense that several peaks in the *HVSR* curves can be observed. For example, the peak at a soil vibration period in the vicinity of 1.5 s is common to all the studies, albeit there are some differences regarding the mobility with period of this peak as a function of the position of the analysed seismic stations. In addition, another peak at vibration periods in excess of

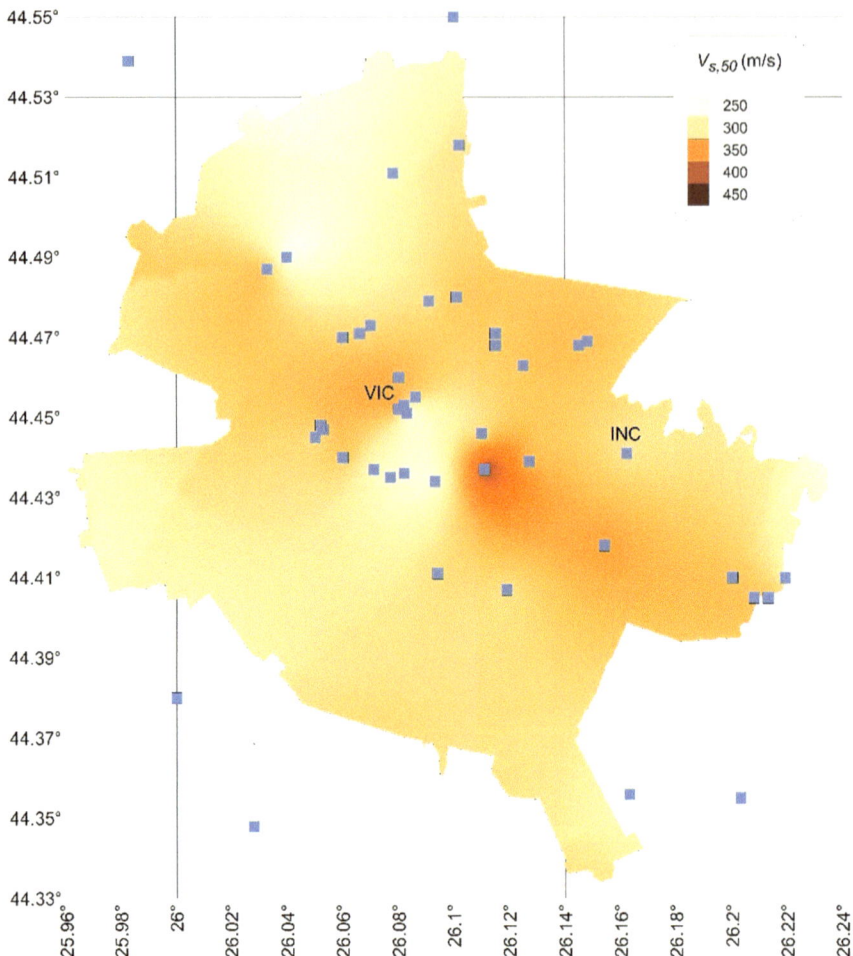

Fig. 2.3 Average shear wave velocities on the first 50 m of soil layers for the available boreholes in the Bucharest area (results from BIGSEES project)

5 s can be inferred from the data shown by Manea et al. (2016). This peak was also noted by Sandi and Borcia (2011) when analysing the transfer function for INCERC site. Manea et al. (2016) note that this peak can be attributed to the interface between Tertiary and Cretaceous (Mesozoic) geological units, while the other peaks can be attributed to the Quaternary soil layers. Some additional peaks below 1 s are visible for some seismic stations within the Bucharest area, as well. Examples of *HVSR* curves for four seismic stations within the Bucharest area are shown in Fig. 2.9.

Manea et al. (2016) conducted an analysis of the waveforms from both ambient vibrations and low-magnitude earthquakes. Thus, the shear-wave velocity profile down to a depth of 8 km was obtained through inversion. Based on the data from

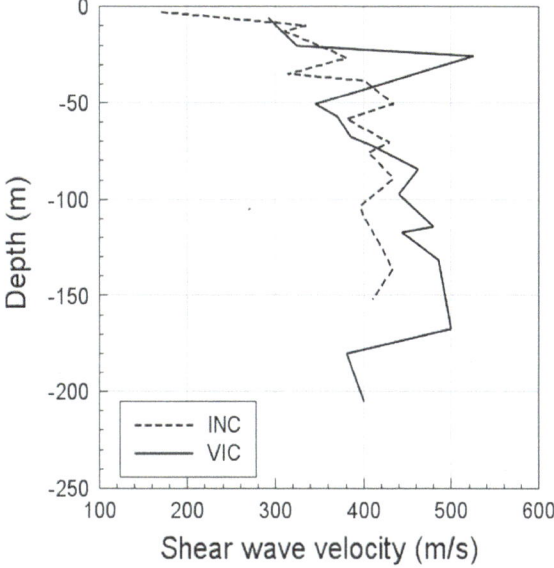

Fig. 2.4 Shear-wave velocity profiles for two deep boreholes in Bucharest

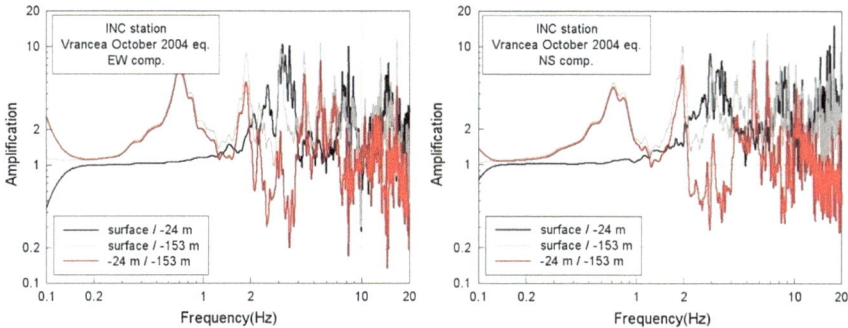

Fig. 2.5 Amplification ratios computed using the two horizontal components recorded at INCERC station during the October 27, 2004 ($M_W = 6.0$, $h = 105$ km) intermediate-depth Vrancea earthquake

Manea et al. (2016), the bedrock is situated at depths in the range 600–1200 m in the Bucharest area, the larger values corresponding to sites in the northern part of the city. Figure 2.10 shows the resulting map for bedrock depth throughout Bucharest city. The proposed map is quite similar to the one proposed by Hartzell (1979). In the next chapter, the influence of this parameter will be evaluated by constructing a ground motion model specifically for the Bucharest area.

The current Romanian seismic design code P100-1/2013 (2013), as well as its previous two versions, define the soil conditions according to the value of the

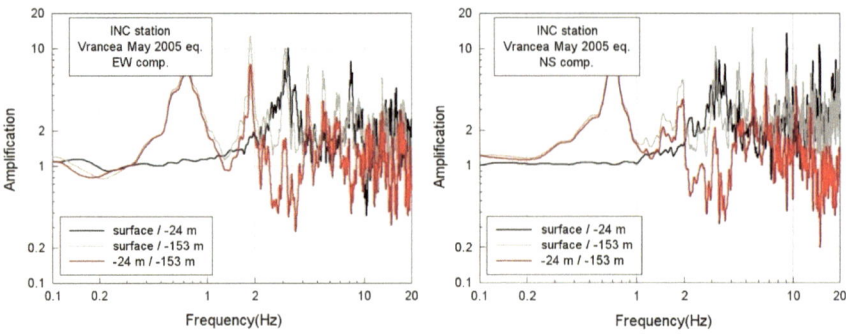

Fig. 2.6 Amplification ratios computed using the two horizontal components recorded at INCERC station during the May 14, 2005 (M_W = 5.5, h = 149 km) intermediate-depth Vrancea earthquake

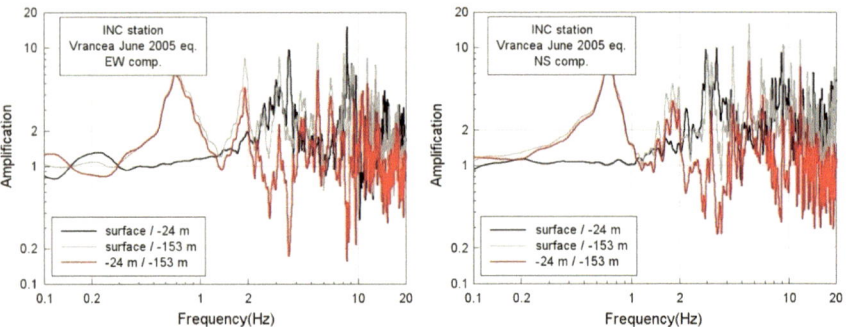

Fig. 2.7 Amplification ratios computed using the two horizontal components recorded at INCERC station during the June 18, 2005 (M_W = 5.2, h = 154 km) intermediate-depth Vrancea earthquake

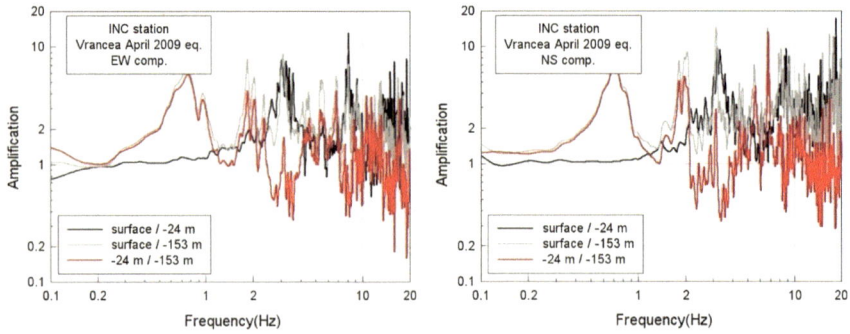

Fig. 2.8 Amplification ratios computed using the two horizontal components recorded at INCERC station during the April 25, 2009 (M_W = 5.4, h = 110 km) intermediate-depth Vrancea earthquake

Fig. 2.9 Examples of mean *HVSR* curves for four seismic stations in Bucharest

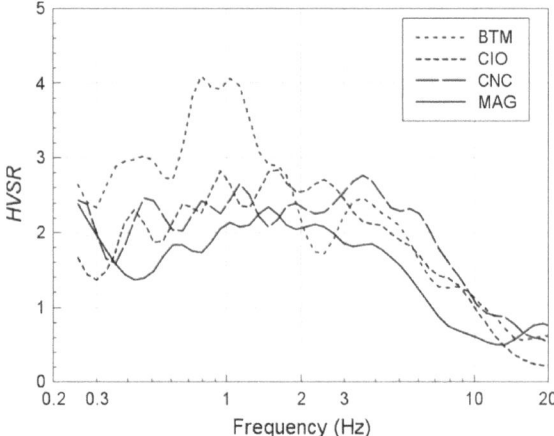

control periods T_C and T_D which represent the border, in the response spectra, between the constant acceleration plateau and the constant velocity plateau and the border between the constant velocity and constant displacement plateau, respectively. The control periods of the acceleration response spectrum defined in the current Romanian seismic design code P100-1/2013 (2013) are obtained using the ground motions recorded during several intermediate-depth Vrancea earthquakes. For each individual ground motion recording, the control periods T_C and T_D are obtained as:

$$T_C = 2\pi \frac{EPV}{EPA} \tag{2.1}$$

$$T_D = 2\pi \frac{EPD}{EPV} \tag{2.2}$$

where *EPA*, *EPV* and *EPD* represent the effective peak acceleration, velocity and displacement. The definitions for *EPA*, *EPV* and *EPD* are based on a sliding/mobile window approach (of 0.4 s width) for averaging the spectral values, as proposed by Lungu et al. (1997).

$$EPA = \frac{\max \overline{SA}_{0.4}}{2.5} \tag{2.3}$$

$$EPV = \frac{\max \overline{SV}_{0.4}}{2.5} \tag{2.4}$$

$$EPD = \frac{\max \overline{SD}_{0.4}}{2.5} \tag{2.5}$$

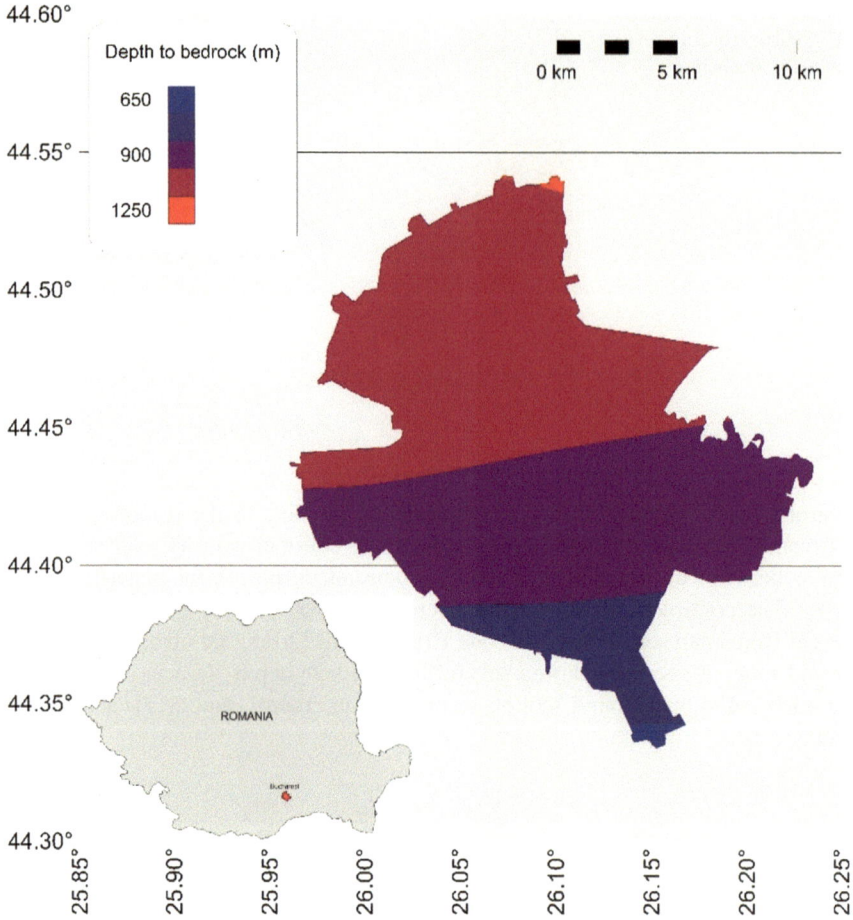

Fig. 2.10 Depth to bedrock in Bucharest based on the results from Manea et al. (2016)

For the Bucharest area, the largest value of T_C of 1, 6 s, assigned by the current Romanian seismic design code, is prescribed. A more detailed analysis of the control periods for both natural and simulated ground motions is given in the next chapter.

Regarding the liquefaction potential of various sites within the Bucharest area, the analyses performed by Arion et al. (2015) show that there are sites within Bucharest that can exhibit liquefaction phenomena during strong intermediate-depth Vrancea earthquakes. These sites are situated near Dambovita River which crosses the city and where recent alluvial deposits, saturated and non-cohesive soils at shallow depths can be encountered.

Pitilakis (2017) proposed an alternative method for deriving the site-specific response spectra for Bucharest by performing 1D nonlinear ground response analysis. Pitilakis (2017) employed two soil profiles in his analysis, the first being the one from the 153 m borehole at INCERC station, also shown in Fig. 2.4, while

Fig. 2.11 Comparison between the spectral acceleration obtained at surface level using nonlinear ground response analysis and equivalent linear ground response analysis

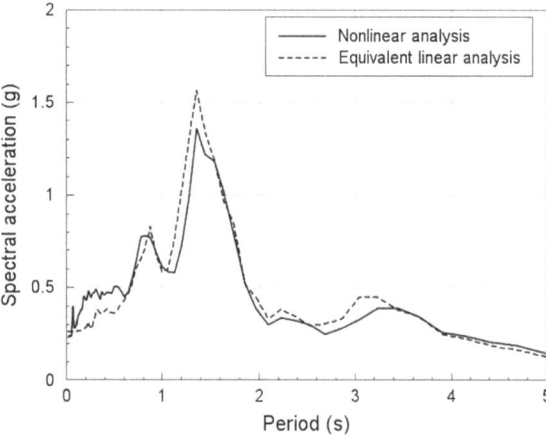

Fig. 2.12 Comparison between the peak ground acceleration profiles with depth obtained using nonlinear ground response analysis and equivalent linear ground response analysis

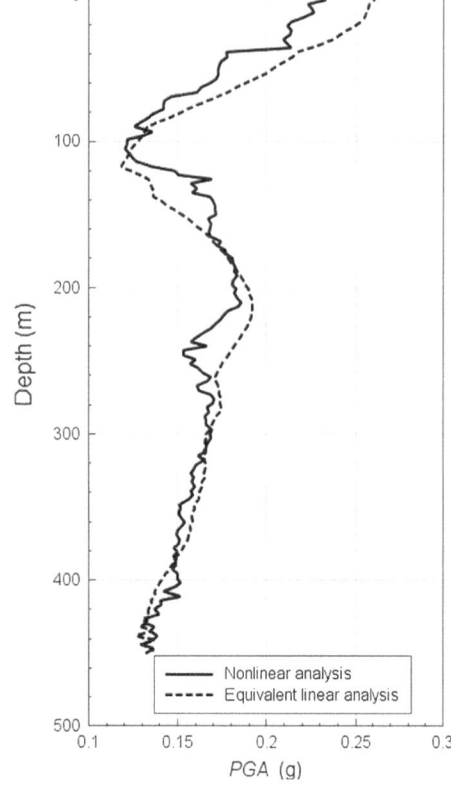

the second is a generic soil profile for Bucharest derived by Cioflan et al. (2009) that extends down to a depth of around 450 m, where the average shear wave velocity is almost 800 m/s, the value considered for the engineering bedrock. In this study, we have applied the same methodology as proposed by Pitilakis (2017), the difference being that we have used reference soil curves derived from Bucharest soil data by Neagu and Arion (2012). The procedure applied in this study also involves the evaluation of the nonlinear soil response starting from a time history of horizontal ground acceleration applied at the bottom of the borehole. The applied time history is based on the deconvolution of the ground motion recorded at INCERC site during the Vrancea earthquake of March 4, 1977 (though a scaling up to a peak ground acceleration of 0.3 g was applied). As shown also in Pitilakis (2017), an equivalent linear analysis was performed, in order to determine the differences between the two approaches for ground response analysis. All the analyses were performed using the software DEEPSOIL (http://deepsoil.cee.illinois.edu/). The response spectra computed at surface level using nonlinear analysis and equivalent linear analysis are shown in Fig. 2.11, while the peak ground acceleration profiles with depth are illustrated in Fig. 2.12. The results are in accordance with the ones obtained by Pitilakis (2017) and show basically larger response spectral values obtained using the equivalent linear approach.

References

Arion C, Calarasu E, Neagu C (2015) Evaluation of Bucharest soil liquefaction potential. Mathe Model Civ Eng 11(1):40–48

Bala A, Hannich D, Ritter JRR, Ciugudean-Toma V (2011) Geological and geophysical model of the quaternary layers based on in situ measurements in Bucharest, Romania. Rom Rep Phys 63:250–274

CEN (2004) Eurocode 8: design of structures for earthquake resistance—part 1: general rules, seismic actions and rules for buildings. European Standard EN 1998-1, Brussels

Cioflan CO, Marmureanu A, Marmureanu G (2009) Nonlinearity in seismic site effects evaluation. Rom J Phys 54:951–963

Grecu B, Radulian M, Mandrescu N (2007) Panza GF (2007) H/V spectral ratios technique application in the city of Bucharest: can we get rid of source effect? JSEE Spring Summer 9:1–14

Hartzell S (1979) Analysis of the Bucharest strong ground motion for the March 4, 1977 Romanian earthquake. B Seismol Soc Am 69:513–530

http://deepsoil.cee.illinois.edu/

Lungu D, Cornea T, Aldea A, Zaicenco A (1997) "Basic representation of seismic action", in design of structures in seismic zones: Eurocode 8—worked examples. In: Lungu D, Mazzolani F, Savidis S (eds) TEMPUS PHARE CM project 01198: implementation of structural Eurocodes in Romanian civil engineering standards. Bridgeman Ltd, Timisoara, pp 1–60

Lungu D, Aldea A, Moldoveanu T, Ciugudean V, Stefanica M (1999) Near-surface geology and dynamic properties of soil layers in Bucahrest. In: Wenzel F, Lungu D (eds) Vrancea earthquakes: tectonics, hazard and risk mitigation. Kluwer Academic Publishers, pp 137–148

Manea EF, Michel C, Poggi V, Fäh D, Radulian M, Balan SF (2016) Improving the shear wave velocity structure beneath Bucharest (Romania) using ambient vibrations. Geophys J Int 207:848–861

Munich Reinsurance Company (ed. 1998), World Map of Natural Hazards, Munich Re, Munich

Nakamura Y (1989) A method for dynamic characteristics estimation of subsurface using microtremor on the ground surface. Q Rep Railway Techn Res Inst 30(1):25–33

Neagu C, Arion C (2012) Dynamic laboratory investigation for soil seismic response. In: Proceedings of the 15th world conference on earthquake engineering, Lisbon, Portugal, paper no. 2051

P100-1/2013 (2013) Code for seismic design—part I—design prescriptions for buildings. Ministry of Regional Development and Public Administration, Bucharest, Romania

Pavel F (2017) Investigation on the variability of simulated and observed ground motions for the Bucharest area. J Earthq Eng. https://doi.org/10.1080/13632469.2017.1297266

Pitilakis K (2017) Site classification and definition of seismic actions in the revision of EC8. In: Keynote lecture presented during the 6th national conference on earthquake engineering and 2nd national conference on earthquake engineering and seismology, Bucharest, Romania

Sandi H, Borcia IS (2011) A summary of instrumental data on the recent strong Vrancea earthquakes, and implications for seismic hazard. Pure appl Geophys 168:659–694

Woessner J, Danciu L, Giardini D, Crowley H, Cotton F, Grünthal G, Valensise G, Arvidsson R, Basili R, Demircioglu MB, Hiemer S, Meletti C, RMW M, Rovida AN, Sesetyan K, Stucchi M, the SHARE Consortium (2015) The 2013 European Seismic Hazard Model: key components and results. Bull Earthq Eng 13(2):3553–3596

Zaharia B, Radulian M, Popa M, Grecu B, Bala A, Tataru D (2008) Estimation of the local response using the Nakamura method for the Bucharest area. Rom Rep Phys 60:131–144

Chapter 3
Characteristics of Simulated and Recorded Ground Motions for Bucharest

Abstract A dataset consisting of more than 400 ground motions recorded during Vrancea intermediate-depth earthquakes that occurred in the period 1977–2013 was compiled in the BIGSEES project financed by the Romanian National Authority for Scientific Research. The analysis of the frequency characteristics of the existing ground motions recorded in the Bucharest area shows that the long-period spectral ordinates increase with the level of the peak ground acceleration, thus denoting possible nonlinear soil effects. The analyses also reveal the fact that the spectral displacement demands imposed by the ground motion recorded during the Vrancea 1977 earthquake at INCERC station shows much larger values as compared to the spectral displacement demands imposed by the ground motions recorded in the Santiago (Chile) area during the Maule 2010 earthquake or in the Wellington (New Zealand) area during the Kaikoura 2016 seismic event. The predictive models proposed for the control periods T_C and T_D highlight the fact that the values of the two periods increase with the earthquake magnitude, while the maximum ratio between spectral acceleration SA and peak ground acceleration PGA decreases with the earthquake magnitude.

Keywords Vrancea intermediate-depth seismic source · Ground motion recordings · Pulse-like ground motions · Spectral displacements
Control periods · Peak ground acceleration · Maximum ratio SA/PGA

Within the BIGSEES research project, a strong ground motion database containing recordings from ten individual earthquakes that occurred in the Vrancea intermediate-depth seismic source in the period 1977–2013 was compiled. The database consists of more than 400 recordings, out of which more than 100 were obtained within the Bucharest city area. Some characteristics of the compiled strong ground motion database are discussed in the paper of Pavel et al. (2014). The mean normalized acceleration response spectra for the ground motions recorded in the Bucharest area during the Vrancea intermediate-depth seismic events of March 1977, August 1986, May 1990 and October 2004 are shown in Fig. 2.1 of the previous section.

© The Author(s) 2018   17
F. Pavel et al., *Impact of Long-Period Ground Motions on Structural Design:
A Case Study for Bucharest, Romania*, SpringerBriefs in Geotechnical
and Earthquake Engineering, https://doi.org/10.1007/978-3-319-73402-6_3

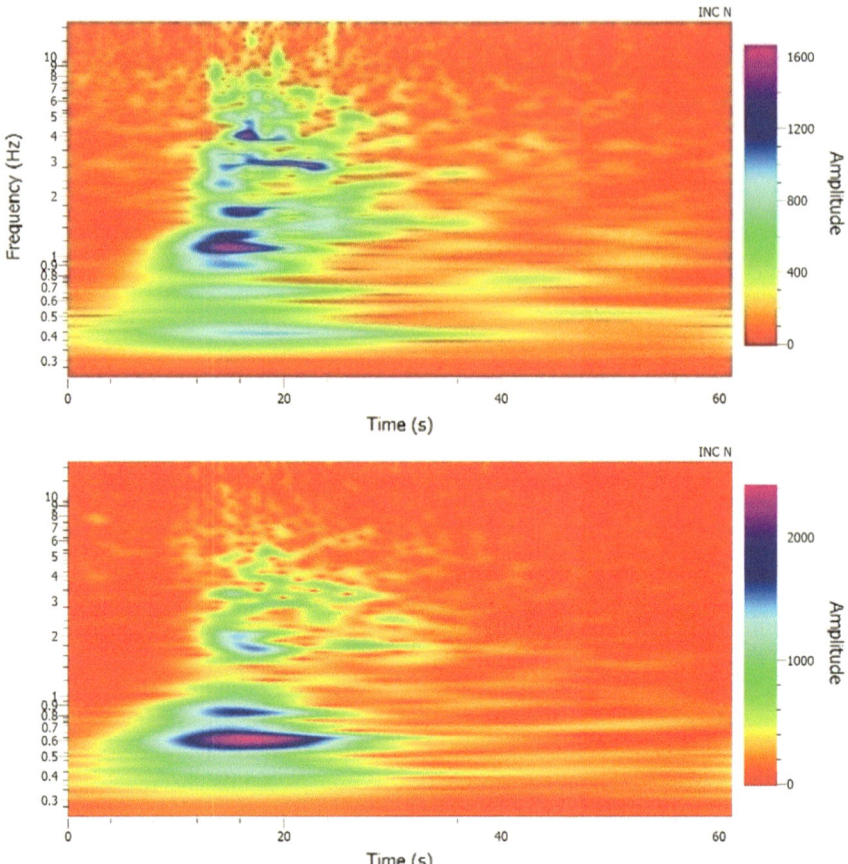

Fig. 3.1 Time-frequency analysis of the two horizontal components recorded at INCERC station during the Vrancea 1977 earthquake (top—EW component, bottom—NS component)

The results of the time-frequency analysis of the two horizontal components recorded at INCERC station during the Vrancea earthquakes of March 1977, August 1986 and May 1990 are shown in Figs. 3.1, 3.2 and 3.3. One can notice that the lowest frequencies (largest periods) were observed during the Vrancea 1977 earthquake, especially on the NS component. The 1986 and the 1990 horizontal components exhibit higher frequencies on both directions. It is also noteworthy that several frequencies are excited within the same time interval of the two recorded horizontal components.

Baker (2006) proposed three criteria in order to characterize a ground motion recording as being pulse-like. The three criteria are as follows:

- the value of a pulse indicator value, defined by Baker (2006), is larger than 0.85;
- the pulse arrives early in the time history;
- the original ground motion has a *PGV* (peak ground velocity) larger than 30 cm/s.

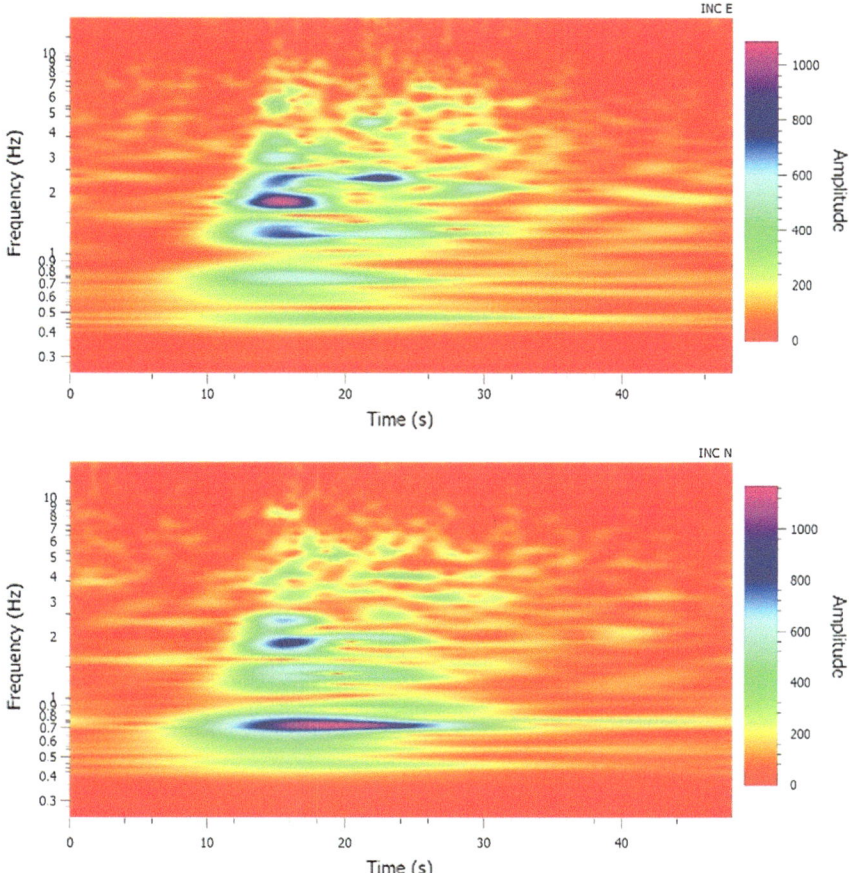

Fig. 3.2 Time-frequency analysis of the two horizontal components recorded at INCERC station during the Vrancea 1986 earthquake (top—EW component, bottom—NS component)

Among all the types of ground motion recorded in the Bucharest area, only two fulfill the three above-mentioned criteria, namely the two horizontal components recorded at INCERC station during the Vrancea intermediate-depth seismic event of March 4, 1977. The value of the pulse indicator is 1.0 for the N-S component and 0.86 for the E-W component. The predominant period of the pulse is 2.58 s for the E-W component and 2.24 s for the N-S component.

The variability of the mean response spectral shape with the corresponding level of peak ground acceleration is analysed in Fig. 3.4. Three levels of peak ground accelerations are considered as follows:

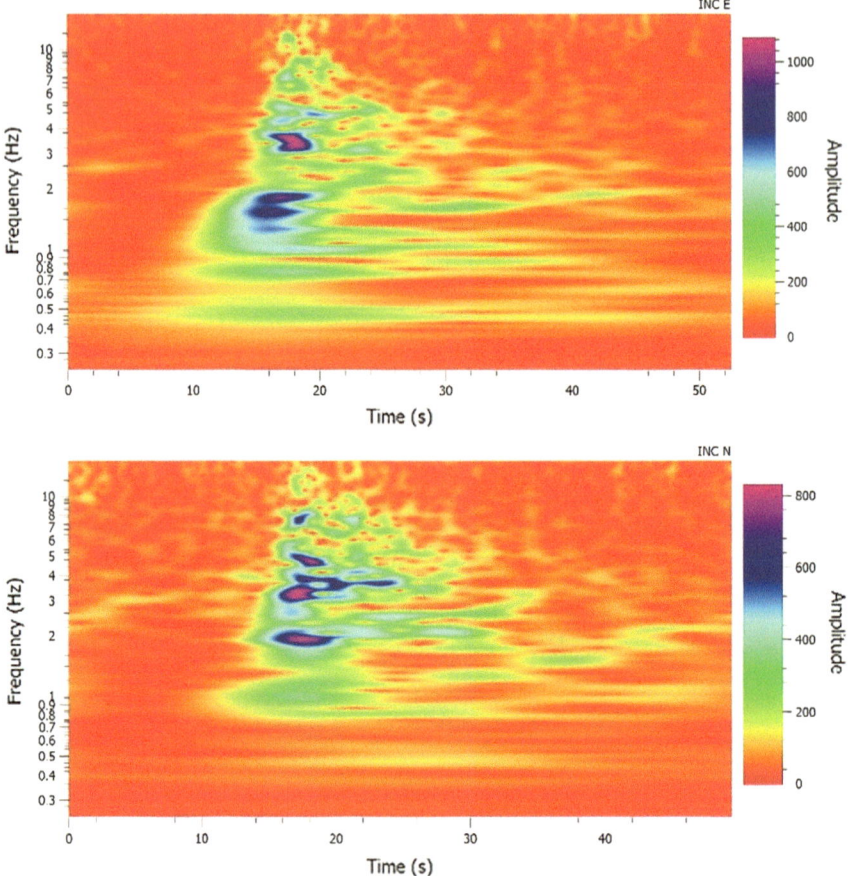

Fig. 3.3 Time-frequency analysis of the two horizontal components recorded at INCERC station during the Vrancea 1990 earthquake (top—EW component, bottom—NS component)

- all the available recordings, irrespective of the values of their peak ground acceleration;
- only ground motion recordings with corresponding values of peak ground accelerations in excess of 0.02 g;
- only ground motion recordings with corresponding values of peak ground accelerations in excess of 0.05 g.

For comparison purposes, the same type of mean normalized response spectra computed using ground motion recorded in Bucharest and in a different part of Romania (the Moldova region), which can be characterized as having a stiffer soil, are also plotted in Fig. 3.4. It is readily apparent that by increasing the level of peak ground acceleration, an increase in the medium- and long-period spectral ordinates can be observed for the recordings in the Bucharest region. On the other hand, this pattern is not visible in the results obtained using ground motions recorded in the

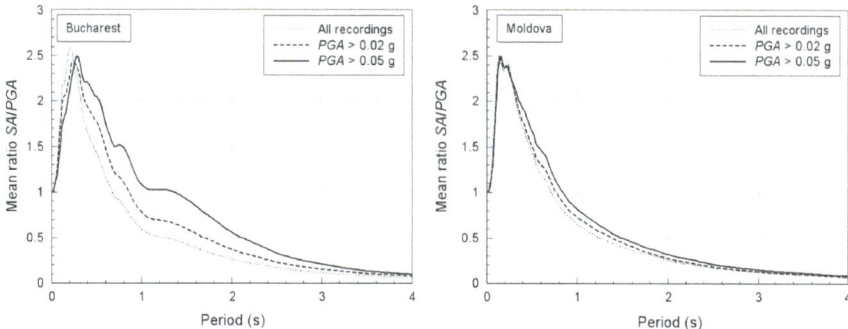

Fig. 3.4 Influence of the peak ground acceleration level on the mean ratio between spectral acceleration and peak ground acceleration for ground motions recorded in the Bucharest area (left) and the Moldova region (right)

Moldova region. Consequently, we can expect significant nonlinear soil effects in Bucharest area during large magnitude Vrancea intermediate depth events.

Another observation with regard to the characteristics of the recorded ground motion (spectral displacement demands) during the Vrancea 1977 earthquake can be inferred from Fig. 3.5. In Fig. 3.5, a comparison of the spectral displacements observed during the Vrancea 1977 seismic event at INCERC station in Bucharest and the spectral displacements observed at five seismic stations within the Santiago (Chile) area during the Maule ($M_W = 8.8$) 2010 earthquake is shown. The recordings from Santiago were chosen based on the fact that the Chilean construction practice resembles the one in Bucharest, namely most of the buildings have a

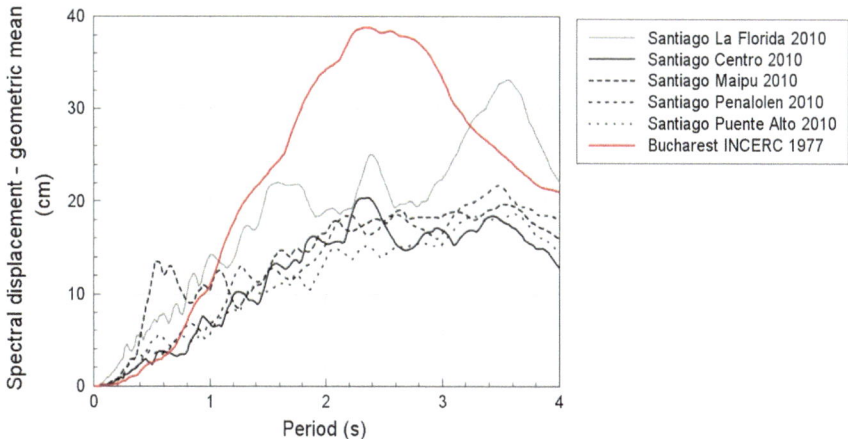

Fig. 3.5 Comparison between the spectral displacements observed at five stations in the Santiago (Chile) area during the Maule 2010 earthquake and the spectral displacements observed at INCERC station during the Vrancea 1977 earthquake

structural system consisting mainly of RC shear walls. The Chilean residential building stock was exposed to the large magnitude 2010 Maule earthquake and suffered limited damage. One of the reasons for the good overall behaviour of high-rise buildings in Chile can be attributed to the limited spectral displacement demands imposed by the ground motions recorded in the Santiago area, even though the corresponding peak ground accelerations were in the range 0.24–0.56 g. Consequently, the much larger (almost double) spectral displacement demands imposed for periods in excess of 1.5 s and up to 3.0 s by the 1977 INCERC recording as compared to the Santiago recordings is highly emphasized.

Another earthquake which has affected a building stock designed primarily using the same principles as those applied for new buildings in Romania was the Kaikoura earthquake that occurred in New Zealand in late 2016. Even though the source-to-site distance between the hypocenter and Wellington was in excess of 200 km, the earthquake caused serious damage to a number of high-rise buildings in this city (even modern buildings). Due to the soft soil conditions beneath Wellington, the Kaikoura earthquake generated considerable spectral displacement demands even though the recorded peak ground acceleration did not exceed 0.21 g. The comparison between the spectral displacements observed at nine stations within the Wellington (New Zealand) area during the 2016 Kaikoura earthquake ($M_W = 7.8$) and the spectral displacements observed at INCERC station during the Vrancea 1977 seismic event is shown in Fig. 3.6. In this case, too, one can notice the much larger spectral displacement demands imposed on high-rise buildings by the ground motion recorded in Bucharest during the Vrancea 1977 earthquake.

In the study of Pavel (2017), more than 3000 stochastic finite-fault simulations were performed for 13 sites in the Bucharest area. The position of the sites is shown in Fig. 3.7. The sites are selected due to the availability of ground motions recorded

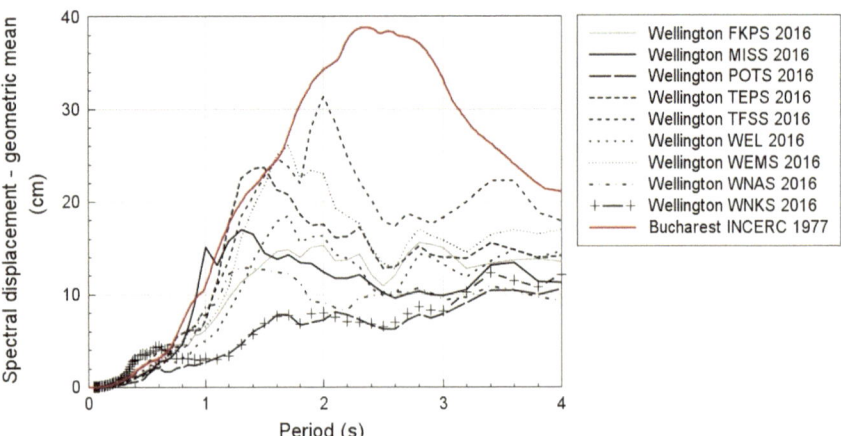

Fig. 3.6 Comparison between the spectral displacements observed at nine stations in the Wellington (New Zealand) area during the Kaikoura 2016 earthquake and the spectral displacements observed at INCERC station during the Vrancea 1977 earthquake

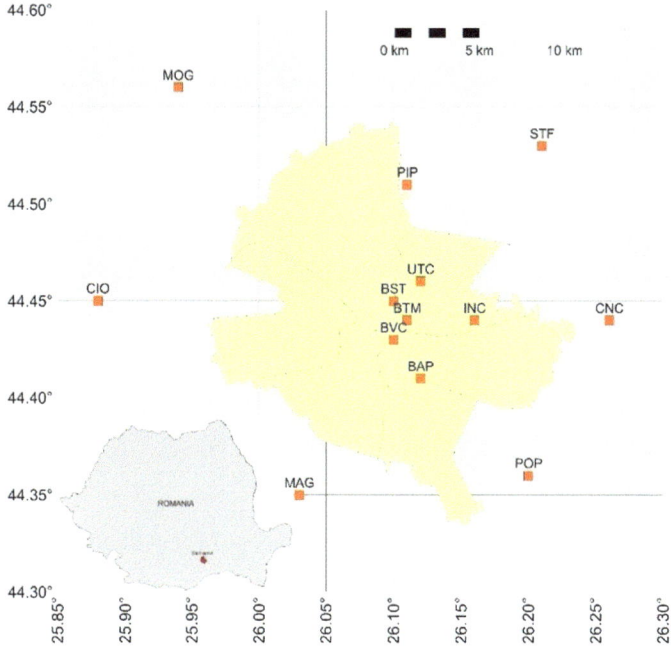

Fig. 3.7 Position of seismic stations for which stochastic finite-fault simulations are performed in the study of Pavel (2017)

during smaller magnitude ($M_W \leq 6.0$) intermediate-depth Vrancea seismic events. The stochastic finite-fault simulations were performed using the EXSIM code (Motazedian and Atkinson 2005) for earthquakes in the magnitude range $M_W = 5.5–7.5$ and focal depths of 90, 110, 130 and 150 km. The positions of the epicenters were proposed based on the seismicity of the XXth century, as shown in the ROMPLUS seismic catalog (http://www.infp.ro/wp-content/uploads/2015/12/romplus.cat_.txt). More details about the simulation procedure can be found in Pavel (2017).

Subsequently, the statistical analysis of some ground motion characteristics, namely the maximum ratio of the spectral acceleration to peak ground acceleration (*PGA*), and the control periods T_C and T_D, is performed for both recorded and simulated ground motions. The results which highlight the influence of the earthquake magnitude and source-to-site distance (in this case, the hypocentral distance) are shown in Figs. 3.8 and 3.9. Two different patterns can be observed in the two previously-mentioned figures. While in the case of the plots showing the influence of the earthquake magnitude, the trends are similar for both recorded and simulated ground motions, the influence of the source-to-site distance highlights a contrasting trend between the two categories of ground motions. The explanation for the two different trends shown in Fig. 3.9 is due to the structure of the database comprising the recorded ground motions. In fact, as shown in Fig. 4.1 of the next chapter, one

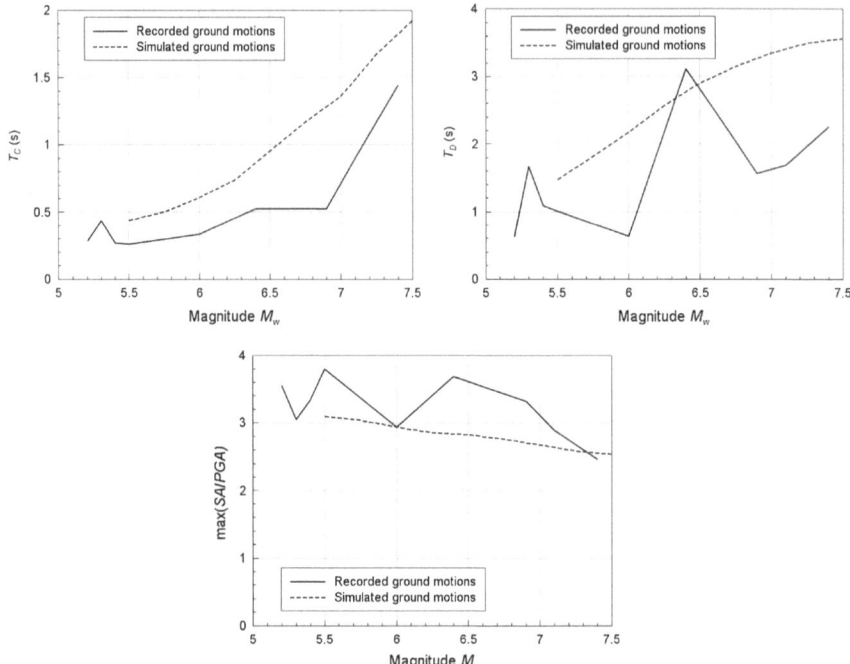

Fig. 3.8 Variation of the control periods T_C, T_D and of the maximum ratio between spectral acceleration and peak ground acceleration as a function of earthquake magnitude for recorded and simulated ground motions

can observe that only small magnitude earthquakes were recorded at large source-to-site distances. As a consequence, smaller values of the control periods T_C and T_D are encountered for the larger source-to-site distances.

Empirical models based on the simulated ground motions are subsequently proposed for three ground motion characteristics, namely the maximum ratio *SA/PGA*, and the control periods T_C and T_D defined in the previous section.

The regression model for the maximum ratio of spectral acceleration to peak ground acceleration [known as maximum dynamic amplification factor in the Romanian seismic design code P100-1/2013 (2013)] has the following functional form, depending only on the earthquake magnitude:

$$\max(SA/PGA) = a + b \cdot M_W + c \cdot M_W^2 + \sigma \qquad (3.1)$$

where the values of the regression coefficients are: $a = 4.8493$, $b = -0.3469$, $c = 0.0051$ and standard deviation $\sigma = 0.2826$.

In the case of the control periods T_C and T_D, the functional form has an additional term reflecting the influence of the source-to-site distance on the two periods. The functional form is given below:

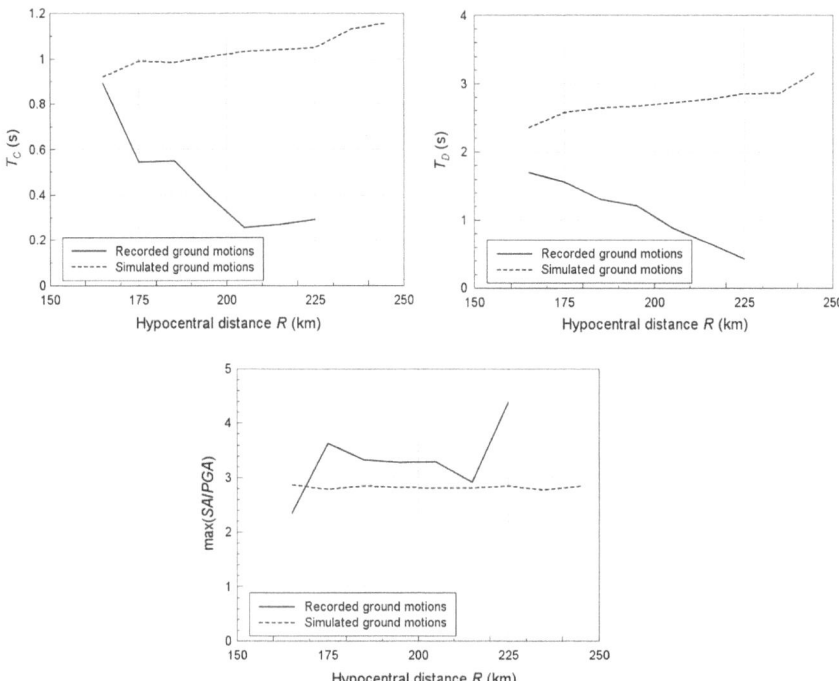

Fig. 3.9 Variation of the control periods T_C, T_D and of the maximum ratio between spectral acceleration and peak ground acceleration as a function of hypocentral distance for recorded and simulated ground motions

$$T_{C,D} = a + b \cdot M_W + c \cdot M_W^2 + d \cdot R + \sigma \qquad (3.2)$$

The values of the regression coefficients are summarized in Table 3.1.

Figure 3.10 shows the variation of the median, median + one standard deviation and median − one standard deviation for the maximum ratio *SA/PGA* given by Eq. (3.1) as a function of the earthquake magnitude. A decreasing trend with the earthquake magnitude is clearly visible in Fig. 3.10. Figure 3.11 analyses the variation of the control periods T_C and T_D obtained using Eq. (3.2) with the earthquake magnitude and for three hypocentral distances, $R = 100$ km, $R = 150$ km and $R = 200$ km. As expected, the largest values of the control periods are observed for the largest

Table 3.1 Regression coefficients for the empirical models for the control periods T_C and T_D

Regression coefficient	T_C	T_D
a	5.9897	−20.6369
b	−2.4218	5.8198
c	0.2447	−0.3637
d	0.0018	0.0056
σ	0.2473	0.4515

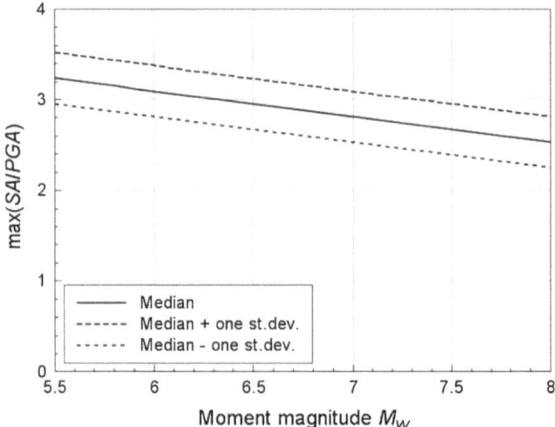

Fig. 3.10 Median, median + one standard deviation and median − one standard deviation values for the maximum ratio *SA/PGA* given by the model in Eq. (3.1)

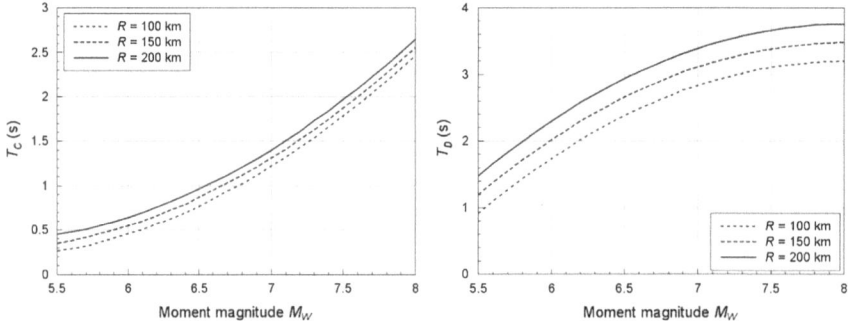

Fig. 3.11 Variation of the control periods T_C and T_D with the earthquake magnitude as a function of the source-to-site distance and using the empirical model given by Eq. (3.2)

hypocentral distance. However, a contrasting trend between the variation of T_C and T_D is observed, in the sense that T_C appears to increase more abruptly as the magnitude increases, while T_D shows a capping value for earthquake magnitudes $M_W \geq 7.5$.

In order to better evaluate the impact of the earthquake magnitude M_W, source-to-site distance R and seismic station soil conditions, ANOVA (analysis of variability) method was applied separately for the recorded and simulated ground motions. The ratio between the spectral accelerations at various spectral periods and the corresponding peak ground acceleration (normalized acceleration response spectra) was considered in the ANOVA analysis. The results in terms of *p*-values are summarized in Tables 3.2 and 3.3. If the computed *p*-values are smaller than 0.05, then the analysed factor is significant. The factors that are not significant are highlighted in bold in Tables 3.2 and 3.3. For both recorded and simulated ground

Table 3.2 p-values for the observed ground motions

Factor	Period T (s)									
	0.2	0.6	1	1.4	1.8	2	2.5	3	3.5	4
M_W	0.000	0.000	0.000	0.000	0.000	0.000	0.000	0.000	0.000	0.000
R	0.000	0.008	0.003	0.000	0.000	0.000	0.000	0.000	0.001	0.033
Station	**0.056**	0.002	0.003	**0.05**	**0.071**	**0.09**	0.018	0.014	0.022	0.229

Table 3.3 p-values for the simulated ground motions

Factor	Period T (s)									
	0.2	0.6	1	1.4	1.8	2	2.5	3	3.5	4
M_W	0.000	0.000	0.000	0.000	0.000	0.000	0.000	0.000	0.000	0.000
R	0.000	0.002	0.000	0.006	**0.532**	**0.405**	**0.321**	**0.135**	0.000	0.000
Station	0.000	0.000	0.000	0.000	0.000	0.000	0.000	0.000	0.000	0.000

motions, the earthquake magnitude has a significant impact on the normalized acceleration response for the entire spectral period range. In the case of the recorded ground motions, the source-to-site distance also influences the normalized acceleration spectral ordinates. On the contrary, the seismic station appears to have little or no influence for this latter category of ground motions. In the case of the simulated ground motions, the seismic station is the second most important parameter after the earthquake magnitude, while the source-to-site distance is the least important.

References

Baker JW (2006) Quantitative classification of near-fault ground motions using wavelet analysis. B Seismol Soc Am 97:1486–1501

http://www.infp.ro/wp-content/uploads/2015/12/romplus.cat_.txt

Motazedian D, Atkinson GM (2005) Stochastic finite-fault modeling based on a dynamic corner frequency. B Seismol Soc Am 95:995–1010

Pavel F (2017) Investigation on the variability of simulated and observed ground motions for the Bucharest area. J Earthq Eng. https://doi.org/10.1080/13632469.2017.1297266

Pavel F, Vacareanu R, Cioflan C, Iancovici M (2014) Spectral characteristics of strong ground motions from intermediate-depth Vrancea seismic source. B Seismol Soc Am 104:2842–2850

Chapter 4
Seismic Hazard Assessment for Bucharest

Abstract In this section, a ground motion model is derived specifically for the Bucharest area using a dataset consisting of both recorded and simulated ground motions from Vrancea intermediate-depth earthquakes. The proposed ground motion model is able to capture the significant long-period spectral ordinates generated by the large magnitude Vrancea intermediate-depth earthquakes occurring at short source-to-site distances. In addition, the spectral accelerations derived from the proposed model have frequency contents similar to the recorded ground motions. The probabilistic seismic hazard assessment performed using the previously mentioned ground motion model shows a plateau of constant spectral accelerations up to around 2.0 s, while the corresponding spectral displacements can reach up to 100 cm for periods of 4.0 s. The analyses also reveal that only the larger magnitude earthquakes occurring at short source-to-site distances can generate spectral ordinates exceeding the ones proposed by the current Romanian seismic design code and this situation is valid mainly for long-period structures.

Keywords Vrancea intermediate-depth seismic source · Ground motion recordings · Ground motion model · Probabilistic seismic hazard assessment Spectral displacements · Hazard curves

The seismic hazard assessment for Romania has been recently conducted within the framework of the SHARE project, financed by the European Union (Woessner et al. 2015) and within the BIGSEES project (Pavel et al. 2016a), funded by the National Authority for Scientific Research in Romania. In both studies, the classic Cornell-McGuire type probabilistic seismic hazard assessment (*PSHA*) was applied using a logic-tree approach with several ground motion prediction equations (*GMPEs*) as branches. The reference soil condition for Bucharest is rock for the SHARE model, and soil class C (according to EN 1998-1/2004, CEN (2004)) for the BIGSEES model.

As previously mentioned, over 100 ground motions were recorded within the Bucharest area during intermediate-depth Vrancea earthquakes occurring in the period 1977–2013. Unfortunately, only one free-field ground motion was recorded

© The Author(s) 2018
F. Pavel et al., *Impact of Long-Period Ground Motions on Structural Design: A Case Study for Bucharest, Romania*, SpringerBriefs in Geotechnical and Earthquake Engineering, https://doi.org/10.1007/978-3-319-73402-6_4

during the large magnitude Vrancea earthquake of March 1977. Therefore, the database containing only natural recordings lacks data from large magnitude intermediate-depth earthquakes. Since it is extremely difficult to find ground motions recorded during similar size intermediate-depth earthquakes in similar soil conditions in other countries, it was decided to augment the existing ground motion database with ground motions simulated using the stochastic finite-fault method, as in Pavel (2017). Out of more than 3000 ground motion simulations from the study by Pavel (2017), it was decided to employ only 100 in order to have an equal share of recorded and simulated ground motions in the final database. In addition, the simulated ground motion recordings were selected so as to cover the magnitude and distance gaps existing in the database of natural ground motion recordings. This aspect is clearly visible in Fig. 4.1 in which the hypocentral distances of both real and simulated ground motion recordings are shown as a function of the earthquake magnitude. The simulated ground motions cover the short and long range of hypocentral distances with only simulated ground motions from earthquakes with $M_W \geq 6.0$ considered in the analysis. The final ground motion database consists of 204 recordings (out of which 104 are actual recordings) from 18 earthquakes (out of which 10 are actual earthquakes). Using the recorded and simulated ground motions previously mentioned, a ground motion prediction equation for Bucharest was developed. The metadata of the earthquakes, as well as the number of the corresponding ground motions used in the regression analysis are given in Table 4.1. The distribution of the peak ground acceleration of the selected recordings as a function of the hypocentral distance is shown in Fig. 4.2 separately for the natural and simulated recordings.

The proposed ground motion model has the following functional form, which is quite similar to the one of the model proposed by Vacareanu et al. (2015):

$$\ln Y = a + b \cdot (M_W - 6) + c \cdot (M_W - 6)^2 + d \cdot h + e \cdot R - \ln R + f \cdot db + \sigma_T \tag{4.1}$$

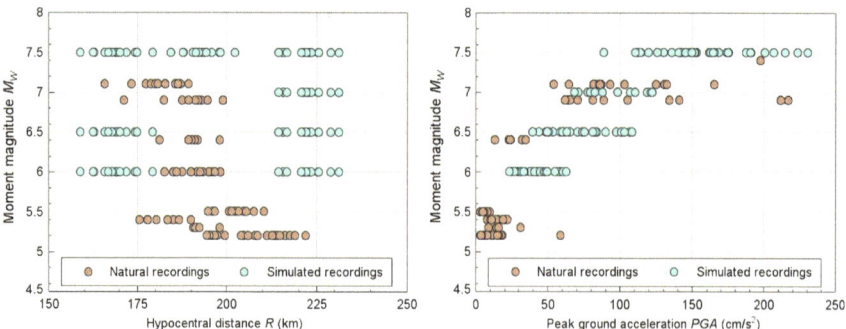

Fig. 4.1 Distribution of the recordings as a function of earthquake magnitude and hypocentral distance (*left*) and as a function of earthquake magnitude and peak ground acceleration (*right*)

Table 4.1 Characteristics of the earthquakes

Earthquake no.	Type	Date	M_W	Focal depth h (km)	No. of recordings
1	Natural	04.03.1977	7.4	94	1
2	Natural	30.08.1986	7.1	131	13
3	Natural	30.05.1990	6.9	91	12
4	Natural	31.05.1990	6.4	87	6
5	Natural	28.04.1999	5.3	151	5
6	Natural	27.10.2004	6.0	105	21
7	Natural	14.05.2005	5.5	149	12
8	Natural	18.06.2005	5.2	154	12
9	Natural	25.04.2009	5.4	110	9
10	Natural	06.10.2013	5.2	135	9
11	Simulated	–	6.0	90	13
12	Simulated	–	6.0	150	13
13	Simulated	–	6.5	90	13
14	Simulated	–	6.5	150	13
15	Simulated	–	7.0	150	13
16	Simulated	–	7.5	90	13
17	Simulated	–	7.5	130	13
18	Simulated	–	7.5	150	13

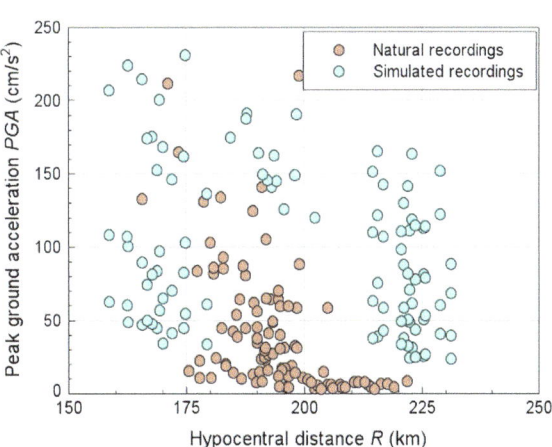

Fig. 4.2 Distribution of the recordings as a function of hypocentral distance and peak ground acceleration

where a, b, c, d, e, f are regression coefficients. Y is the geometric mean of the two horizontal components for a given spectral period T, in cm/s^2, M_W is the earthquake moment magnitude, h is the earthquake focal depth, R is the source-to-site distance (in this case hypocentral distance) and db is the depth to bedrock. All distances are measured in km. The bedrock depth db is inferred for each seismic station from the map shown in Fig. 2.10. The total standard deviation of the model σ_T is composed of both intra- and inter-event variability (denoted as σ and τ) and is computed as:

$$\sigma_T = \sqrt{\sigma^2 + \tau^2} \tag{4.2}$$

where σ is the intra-event variability and τ is the inter-event variability.

The values of the regression coefficients are given in Table 4.2. Referring to the last three columns of the table, one can notice that the largest part of the variability is due to the intra-event variability for short spectral periods, while for larger spectral periods, both the inter- and intra-event variability have almost equal shares. The regression model also shows that the depth to bedrock acts in a similar manner for almost the entire period range in the sense that it decreases the ground motion amplitudes for larger depths. This trend is especially visible for periods that are close to those of the various soil layers underneath Bucharest.

Figure 4.3 compares the attenuation with distance (epicentral distance) of the proposed model and of the Vacareanu et al. (2015) model (VEA15), for two pairs of magnitudes ($M_W = 6.5$ and 7.5) and focal depths ($h = 90$ and 150 km). The comparison highlights a decrease in the difference between the median amplitudes of the two ground motion models as the earthquake magnitude is increasing. In addition, the proposed model has a smaller variability than the Vacareanu et al. (2015) model due to a better constrained ground motion database (only ground motions recorded in the Bucharest area are employed). However, one has to take into account the fact that the proposed model is valid only for the Bucharest region and cannot be extrapolated for other regions, while the Vacareanu et al. (2015) model can be applied at regional scale.

Table 4.2 Regression coefficients for the proposed ground motion model

$T(s)$	a	b	c	d	e	f	σ_T	σ	τ
0.0	9.5939	1.3873	−0.2249	0.0018	−0.0055	−0.0454	0.3375	0.2950	0.1640
0.2	10.1819	1.1954	−0.1931	−0.0001	−0.0022	−0.1327	0.3581	0.3256	0.1491
0.4	10.0232	1.7910	−0.4303	0.0034	−0.0060	0.0355	0.3915	0.3284	0.2132
0.6	9.6068	2.0506	−0.5296	0.0030	−0.0049	−0.0304	0.3353	0.2763	0.1900
0.8	10.4373	2.3478	−0.6625	0.0111	−0.0145	−0.2336	0.3002	0.2362	0.1853
1.0	9.6985	2.4467	−0.6252	0.0102	−0.0130	−0.0131	0.2897	0.2226	0.1855
1.2	9.8940	2.6353	−0.6719	0.0143	−0.0169	−0.1572	0.2963	0.2275	0.1899
1.4	9.8487	2.8010	−0.7289	0.0168	−0.0186	−0.2140	0.3187	0.2491	0.1987
1.6	9.8158	2.9278	−0.7618	0.0189	−0.0202	−0.2939	0.3271	0.2557	0.2040
1.8	9.3450	3.0085	−0.7419	0.0189	−0.0193	−0.1781	0.3368	0.2612	0.2126
2.0	8.7276	3.0504	−0.7151	0.0175	−0.0166	−0.0854	0.3516	0.2734	0.2211
2.5	7.6764	3.0441	−0.6244	0.0130	−0.0107	−0.0062	0.3763	0.2982	0.2295
3.0	6.9975	3.1449	−0.6038	0.0106	−0.0069	−0.1429	0.4300	0.3420	0.2607
3.5	6.6436	3.1580	−0.5432	0.0100	−0.0050	−0.3987	0.4505	0.3552	0.2771
4.0	6.1576	3.2014	−0.5292	0.0087	−0.0023	−0.4787	0.4794	0.3761	0.2973

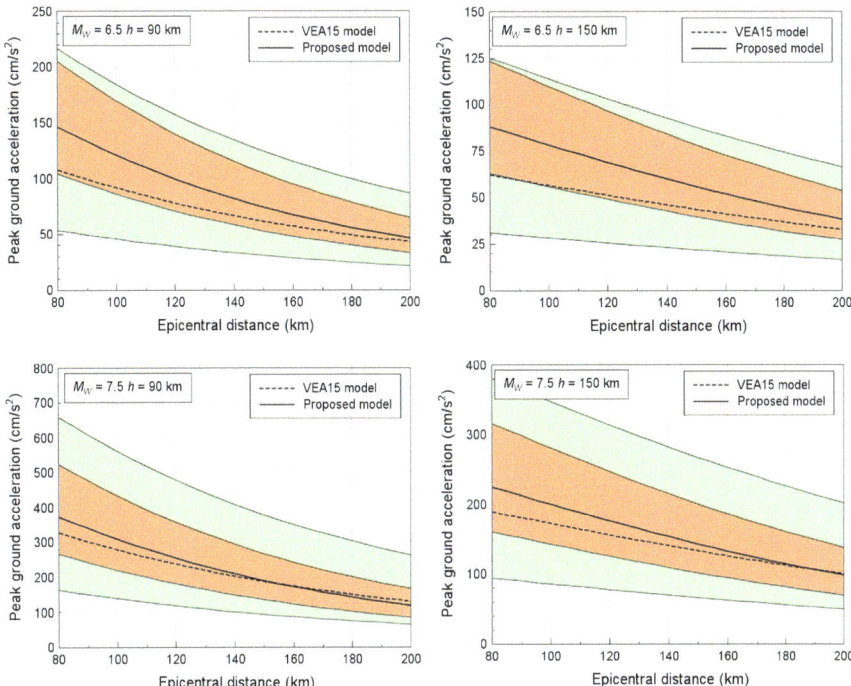

Fig. 4.3 Comparison of distance attenuation for the proposed ground motion model and for the Vacareanu et al. (2015)—VEA15 model for four earthquake scenarios. The colored range represents the area between the median—one standard deviation curve and median + one standard deviation curve for the two models

In Fig. 4.4 the response spectra for earthquakes having $M_W = 6.5$ and 7.5 and source-to-site distances of 135 and 225 km are shown. The effect of the source-to-site distance on the long-period spectral amplitudes is clearly visible in Fig. 4.4. Even though only larger magnitude earthquakes are capable of generating significant long-period spectral ordinates, it is clear that short source-to-site distances will significantly increase the spectral ordinates in the range 0.5–1.5 s.

A comparison between the acceleration response spectra observed at INCERC station during the Vrancea intermediate-depth seismic events of March 1977, August 1986 and May 1990 and the median spectra obtained using the proposed ground motion model is presented in Fig. 4.5.

Figure 4.5 shows that the proposed ground motion model is able to capture the significant long-period spectral ordinates observed at INCERC station during the Vrancea earthquakes of March 1977 and August 1986 and the lack of significant long-period spectral ordinates observed during the Vrancea seismic event of May 1990.

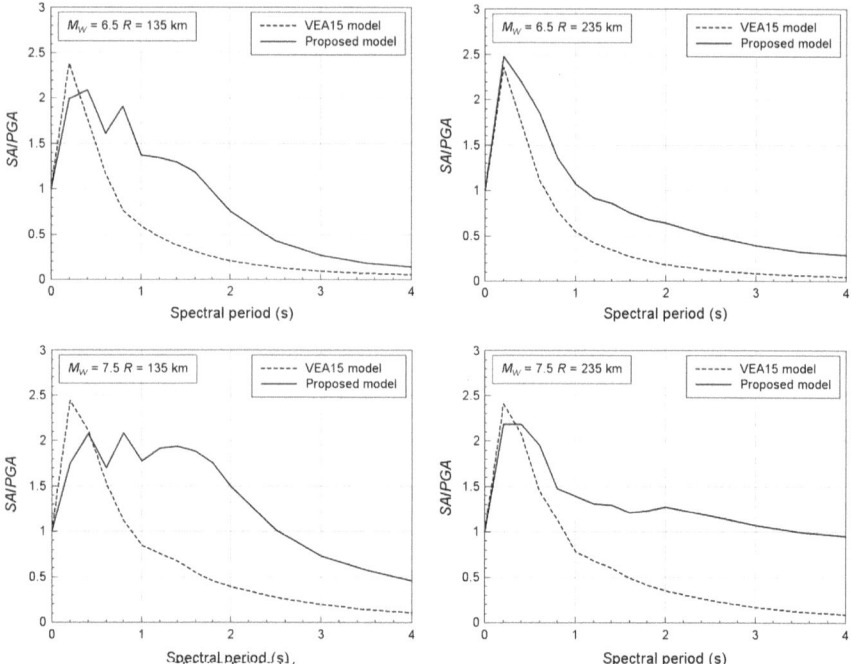

Fig. 4.4 Comparison of response spectra obtained using the proposed ground motion model and using the Vacareanu et al. (2015)—VEA15 model for four earthquake scenarios

Subsequently, a probabilistic seismic hazard assessment was performed for Bucharest by employing the purposely defined ground motion model. The sites considered in the analysis are shown in Fig. 4.6. Only the Vrancea intermediate-depth seismic source was considered in the analysis. The truncation limit of the lognormal distribution of ground motion amplitudes was considered at three standard deviations ($\varepsilon = 3$).

The uniform hazard spectra (*UHS*) corresponding to a mean return period of 475 years for the points situated at the extremities of Bucharest, namely points 1, 3, 8 and 9 are shown below. The results were obtained using the ground motion model purposely developed for Bucharest. For comparison purposes, the uniform hazard spectrum obtained using the ground motion model VEA15 is also shown. The results for spectral accelerations are given in Fig. 4.7, while the results for spectral displacements are shown in Fig. 4.8. Figures 4.7 and 4.8 show that the differences among the four selected points are larger in terms of spectral accelerations than in terms of spectral displacements.

Pavel et al. (2016b) evaluated the seismic hazard for INCERC site in the eastern part of Bucharest using stochastic finite-fault simulations performed using a Monte Carlo simulated earthquake catalogue for the Vrancea intermediate-depth seismic source. Pavel and Vacareanu (2017) applied the classic Cornell-McGuire type

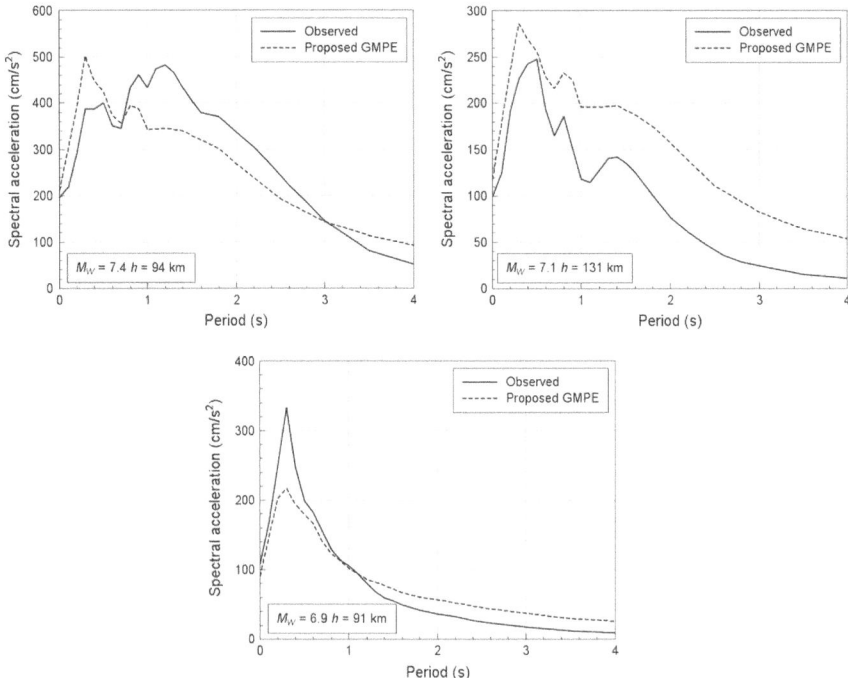

Fig. 4.5 Comparison between the observed response spectra and the response spectra derived using the proposed ground motion model for INCERC seismic station and for the Vrancea intermediate-depth earthquakes of March 1977, August 1986 and May 1990

probabilistic seismic hazard assessment with the consideration of site-dependent soil amplifications derived from stochastic finite-fault simulations. Both approaches highlighted the high spectral ordinates corresponding to medium and long spectral periods. The results from the above-mentioned two studies are compared with the results obtained in this chapter in Figs. 4.9, 4.10 and 4.11. In addition, the results from Pavel et al. (2016a) are also shown for comparison purposes, denoted as classic *PSHA*.

Figures 4.9, 4.10 and 4.11 show significant differences between the four seismic hazard studies, especially for $T = 1.0$ s and $T = 2.0$ s. It is noteworthy that in the range of exceedance probabilities used for seismic design, the values of peak ground acceleration are quite similar, irrespective of the study.

Figure 4.12 compares the elastic spectral displacements for Bucharest given by the current Romanian seismic design code and the median spectral displacements predicted by the proposed ground motion model for three magnitudes $M_W = 7.0$, $M_W = 7.4$ and $M_W = 7.8$, and two source-to-site distances (hypocentral distance) $R = 135$ km and $R = 215$ km. The results highlight two main features, namely the fact that the code spectral displacements for vibration periods in excess of 2.0 s are inferior to the median ones predicted by the ground motion model for earthquakes

Fig. 4.6 Distribution of
points for which the
assessment of seismic hazard
was performed

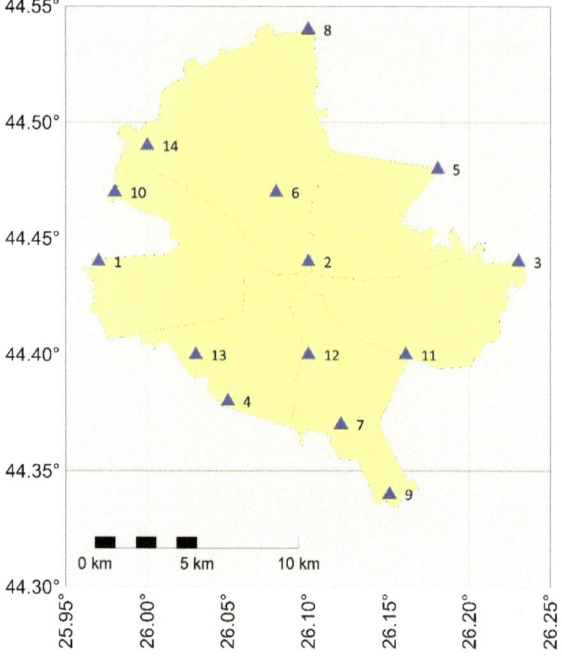

Fig. 4.7 Comparison of
uniform hazard spectra
(spectral accelerations) for
four points in Bucharest
evaluated using the proposed
ground motion model and the
uniform hazard spectra
determined using the VEA15
model

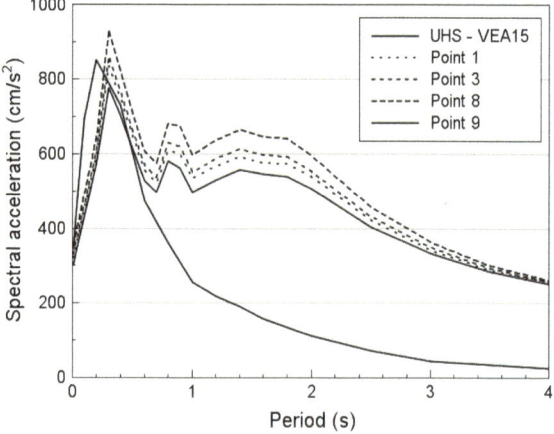

of magnitude $M_W = 7.8$, and the fact that the control period T_D increases with the
earthquake magnitude and source-to-site distance. This conclusion is similar to the
one observed in Chap. 3 when deriving the regression model for the control period
T_D. However, it is apparent from Fig. 4.12 that the control periods T_D are in excess
of 4 s.

Fig. 4.8 Comparison of
uniform hazard spectra
(spectral displacements) for
four points in Bucharest
evaluated using the proposed
ground motion model and the
uniform hazard spectra
determined using the VEA15
model

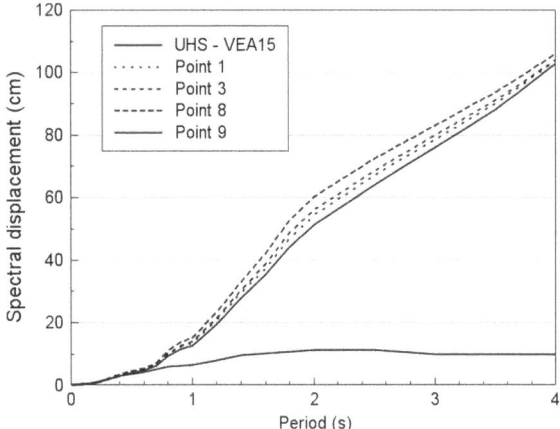

Fig. 4.9 Comparison of
hazard curves for peak ground
acceleration obtained in four
seismic hazard studies

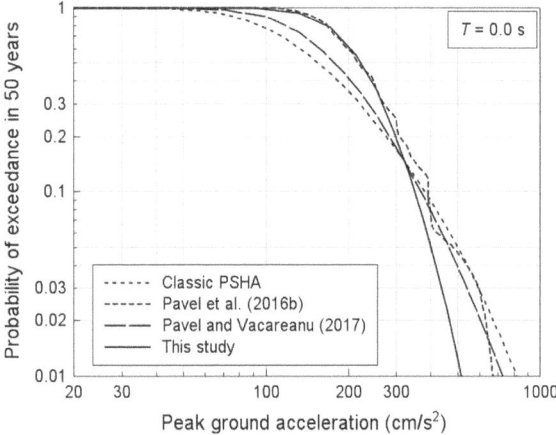

Fig. 4.10 Comparison of
hazard curves for $T = 1.0$ s
obtained in four seismic
hazard studies

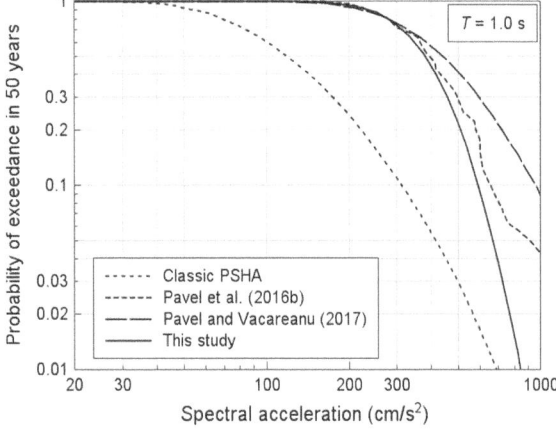

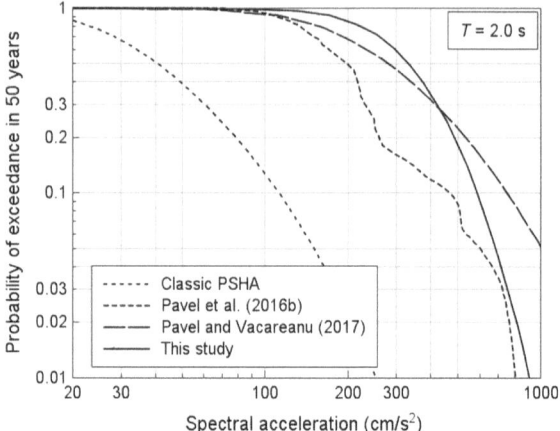

Fig. 4.11 Comparison of hazard curves for $T = 2.0$ s obtained in four seismic hazard studies

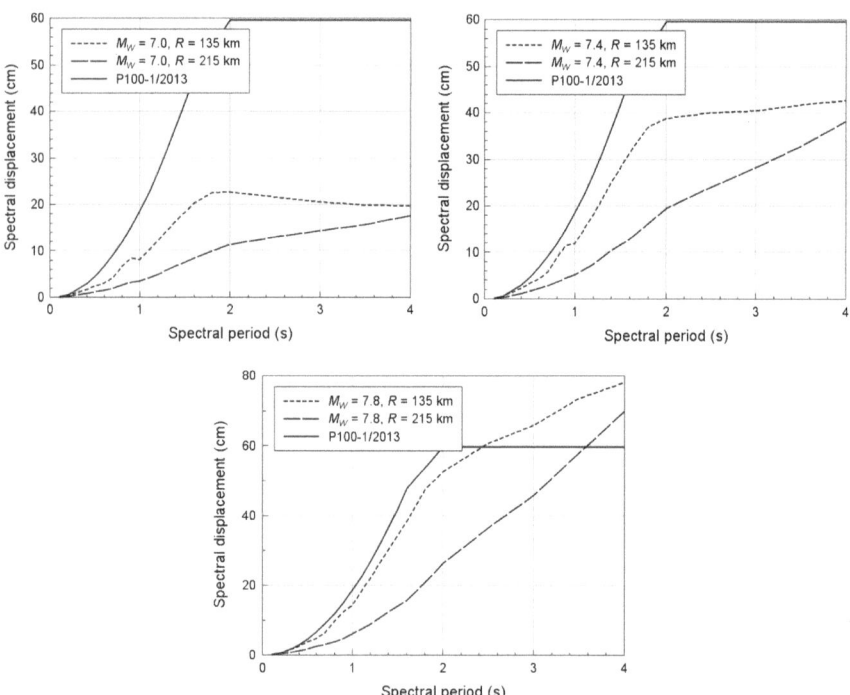

Fig. 4.12 Comparison between the median spectral displacements predicted by the proposed ground motion model for various magnitudes and source-to-site distances and the elastic spectral displacements given in the Romanian seismic design code P100-1/2013 (2013)

References

CEN (2004) Eurocode 8: design of structures for earthquake resistance—part 1: general rules, seismic actions and rules for buildings. European Standard EN 1998-1, Brussels

P100-1/2013 (2013) Code for seismic design—part I—design prescriptions for buildings. Ministry of Regional Development and Public Administration, Bucharest, Romania

Pavel F (2017) Investigation on the variability of simulated and observed ground motions for Bucharest area. J Earthq Eng. https://doi.org/10.1080/13632469.2017.1297266

Pavel F, Vacareanu R, Douglas J et al (2016a) An updated probabilistic seismic hazard assessment for Romania and comparison with the approach and outcomes of the SHARE project. Pure Appl Geophys 173:1881–1905

Pavel F, Ciuiu D, Vacareanu R (2016b) Site dependent seismic hazard assessment for Bucharest based on stochastic simulations. In: Vacareanu R, Ionescu C (eds) The 1940 Vrancea earthquake. Issues, insights and lessons learnt. Springer natural hazards. Switzerland, pp 221–233

Pavel F, Vacareanu R (2017) Ground motion simulations for seismic stations in sourthern and eastern Romania and seismic hazard assesment. J seismol 21:1023–1037

Vacareanu R, Radulian M, Iancovici M, Pavel F, Neagu C (2015) Fore-arc and back-arc ground motion prediction model for Vrancea intermediate depth seismic source. J Earthq Eng 19:535–562

Woessner J, Danciu L, Giardini D et al (2015) The 2013 European seismic hazard model: key components and results. B Earthq Eng 13:3553–3596

Chapter 5
Structural Design for Large Displacement Demands in Romania

Abstract In this section, some aspects regarding the structural design and detailing of reinforced concrete buildings in Bucharest are discussed. The evaluation of the spectral shapes used in the design of buildings is discussed, as well. A new relation for the evaluation of the amplification factor for spectral displacements is proposed based on recent results obtained using ground motions recorded during intermediate-depth Vrancea earthquakes. Examples of the main structural systems used for common building in Bucharest are given. The issue of designing structural systems for relatively high levels of lateral force coupled with large displacement demands is also discussed in depth. As a general observation, the shear wall density for buildings in Bucharest is almost twice as large as that of Chilean buildings. An illustrative example for the structural design of a medium-rise reinforced concrete building in Bucharest is given and the main issues regarding the structural detailing are highlighted.

Keywords Seismic design code · Spectral displacements · Ductility class
Behaviour factor · Structural systems · Limit states · Storey drift

5.1 Basics of Structural Design

The national code for seismic design of buildings currently enforced in Romania is P100-1/2013 (2013). This code is imposed by the Romanian Government and is compulsory for the entire Romanian territory. A brief overview of the Romanian seismic design code, P100-1, is presented by Popa (2014). P100-1/2013 (2013) is harmonized with EN 1998-1 (2004b) with regards to design procedure, format, symbols and definitions and contains additional specific recommendations for Romania. These are mainly related to the definition and the calibration of the design seismic action, capacity design procedure and detailing rules. The entire design process is based on capacity design principles.

© The Author(s) 2018   41
F. Pavel et al., *Impact of Long-Period Ground Motions on Structural Design:
A Case Study for Bucharest, Romania*, SpringerBriefs in Geotechnical
and Earthquake Engineering, https://doi.org/10.1007/978-3-319-73402-6_5

Fig. 5.1 Fundamental
requirements for seismic
design in P100-1/2013 (2013)

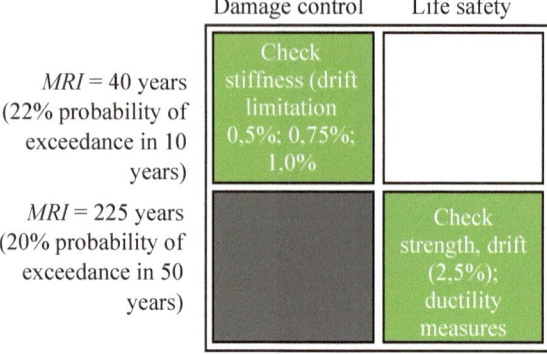

The chart shows:

Damage control | Life safety

MRI = 40 years
(22% probability of
exceedance in 10
years) — Check stiffness (drift limitation 0.5%; 0.75%; 1.0%)

MRI = 225 years
(20% probability of
exceedance in 50
years) — Check strength, drift (2.5%); ductility measures

Similar to EN 1998-1 (2004b), P100-1/2013 (2013) is based on a performance-based approach with two fundamental requirements: damage control and life safety (Fig. 5.1).

The evolution of the response spectral shapes used for the design of new structures in Romania is shown in Fig. 5.2. Briefly, the seismic codes can be divided into two categories: codes enforced before the 1977 Vrancea earthquake—P13/63 and P13/70, and codes enforced after the 1977 seismic event—P100/78 and P100/81, P100/92, P100-1/2006 and P100-1/2013. In terms of peak ground accelerations used for the design of buildings in Bucharest, the values increased from 0.05 g (or 0.03 g) before 1977 to 0.20 g immediately after the 1977 earthquake, and then to 0.24 g in the 2006 version of the code (probability of exceedance of 39% in 50 years); in the 2013 version the value increased again to 0.30 g (probability of exceedance of 20% in 50 years). The reader can easily observe the very large difference between the code-based spectral shapes for Bucharest before and after 1977 for vibration periods in excess of 0.5 s. The value of the soil factor S of EN 1998-1 (2004b) can be assumed as 1, as shown in the study of Vacareanu et al. (2014).

Fig. 5.2 Evolution of code
response spectral shapes for
Bucharest

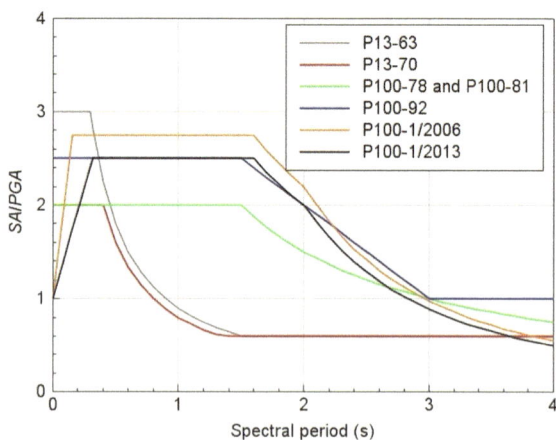

For damage control, the design process is focused on ensuring enough stiffness to prevent large lateral deformation under a seismic event with 22% probability of exceedance in 10 years (service earthquake). This implies the limiting of the drift angle to 0.5, 0.75 or 1.0%, depending on the type of non-structural components. For masonry infilled walls the first limit is prescribed, while for non-structural components completely detached from the main structures the later limit can be observed. Limited nonlinear response of the reinforced concrete structures is allowed. For the fundamental life safety requirement, the design process focuses on lateral strength, drift limitation and ductility to prevent loss of life or irreparable structural damage for a seismic event with 20% probability of exceedance in 50 years (design earthquake).

For structures with fundamental vibration periods lower than the control period T_C, a displacement amplification factor is given in the Romanian seismic design code P100-1/2013 (2013). This displacement amplification factor takes into account the fact that the inelastic displacements of structures with fundamental vibration periods lower than that of the control period T_C are larger than the corresponding elastic displacements. The relation given in the Romanian seismic code is:

$$1 \leq c = 3 - 2.3 \cdot \frac{T}{T_C} < \frac{\sqrt{T_C \cdot q}}{1.7} \tag{5.1}$$

where q is the behaviour factor used for structural design and T is the fundamental period of the structural model. The ground motion database that was used in the regression analysis for the calibration of the coefficients consisted only of artificial spectrum-compatible time histories.

Recently, Craciun et al. (2016) have proposed another relation with the following functional form:

$$1 \leq c = 0.25 + 0.60 \cdot \frac{T_C}{T} + 0.5 \cdot \ln q \leq 2 \tag{5.2}$$

An upper bound equal to two, similar to the one given in the previous version of the Romanian seismic design code (2006 version), was introduced in the relation. The database employed for the regression consisted of ground motions recorded during the Vrancea intermediate-depth earthquakes of August 1986 ($M_W = 7.1$, $h = 131$ km) and May 1990 ($M_W = 6.9$, $h = 94$ km and $M_W = 6.4$, $h = 87$ km). The analyses were performed using the well-known Takeda hysteretic model for behaviour factors q in the range 1.0–6.0. Figure 5.3 compares the displacement amplification factors obtained by applying the relation proposed by the seismic design code P100-1/2013 (2013) and the amplification factors computed using relation (5.2), for two values of the behaviour factor q. One can observe from Fig. 5.3 that there is a significant difference between the displacement amplification factors obtained by applying the two relations in the sense that the relation derived by Craciun et al. (2016) gives much larger values. In addition, it is clear that the

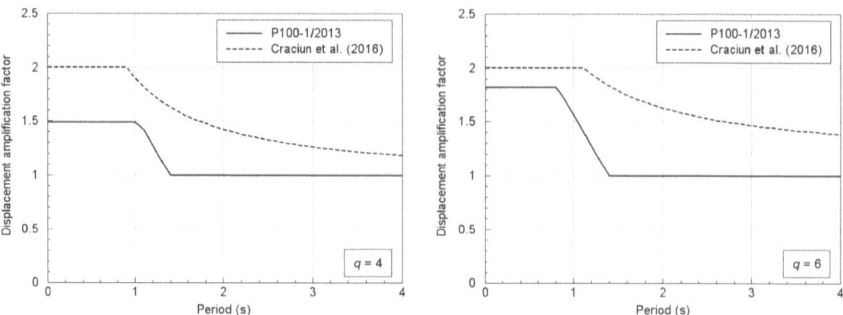

Fig. 5.3 Comparison between the displacement amplification factors for Bucharest given by the current Romanian seismic design code P100-1/2013 and those proposed by Craciun et al. (2016) as a function of the behaviour factor q

difference between the two sets of displacement amplification factors increases with the value of the behaviour factor q.

The displacement amplification factors proposed by Craciun et al. (2016) are combined with the spectral displacements obtained using the ground motion model proposed in the previous chapter. The results in terms of inelastic spectral displacements for several earthquake scenarios are shown in Figs. 5.4 and 5.5, as a function of the behaviour factor q. Among the most important re observations drawn from Figs. 5.4 and 5.5 is the fact that even in the case of inelastic spectral displacements, there is a net increase of the predominant period of the response spectrum as a function of the earthquake magnitude for short source-to-site distances. However, in the case of distant earthquakes, the inelastic spectral displacements increase gradually with the spectral period. It is clear that for spectral periods T in excess of 4.0 s, the inelastic spectral displacements will tend to become equal for both types of earthquakes with short and long source-to-site distances. As expected, more distant earthquakes will tend to have a greater effect on taller structures. The very large discrepancy in spectral displacements for periods in the range 1.0–3.0 s is also noteworthy.

In order to validate the previously computed inelastic spectral displacements, the observed values and predicted values for the INCERC seismic station during the Vrancea earthquake of March 1977 are compared in Fig. 5.6. The similarity in spectral displacements for both $q = 4$ and $q = 6$ both up to $T = 2.5$ s is noteworthy. For periods in excess of 2.5 s, there are considerable differences between observed and predicted inelastic spectral displacements; however, one has to take into account that the processing of the original strong ground motion recorded at INCERC station was performed using a low-cut filter of 0.25 Hz, thus the reliability of the spectral displacements for periods in excess of 2.5–3.0 s is questionable.

Similar to EN 1998-1 (2004b), P100-1/2013 (2013) allows the design for three ductility classes: high (DCH), medium (DCM) and low (DCL). The implicit

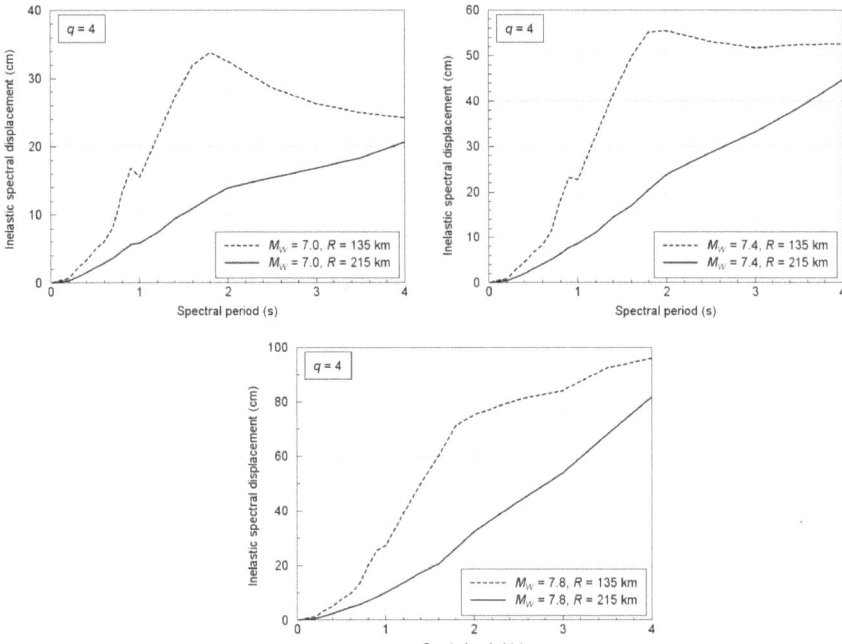

Fig. 5.4 Comparison between the median inelastic spectral displacements function of the earthquake magnitude and source-to-site distance for a behaviour factor $q = 4$

ductility class prescribed by P100-1 is DCH, as opposed to EN 1998-1 (2004b) which prescribes DCM. According to P100-1, DCL is allowed only in low seismicity areas (design peak ground acceleration smaller than 0.1 g) and cannot be used for most of the national territory. Behaviour factors for concrete buildings in P100-1 are in general 10% larger than those prescribed by EN 1998-1 (2004b). The ranges of the behaviour factors from the Romanian seismic design code for the three ductility classes are given in Table 5.1.

To tailor the seismic requirements on the consequence of failure, P100-1/2013 (2013) classifies the structures into four classes of importance. This classification largely follows the provisions of ASCE 7 standard. In addition, P100-1 classifies the buildings over 45 m in height as importance class I and buildings over 28 m (but up to 45 m) as importance class II. For importance class I, a 40% increase in the design spectral ordinates with respect to the values for normal importance buildings (importance class III) is prescribed. For importance class II, a 20% increase is prescribed. This provision is implemented in the design code to restrain the construction of mid-to high-rise buildings that are most sensitive to seismic action with long predominant spectral periods.

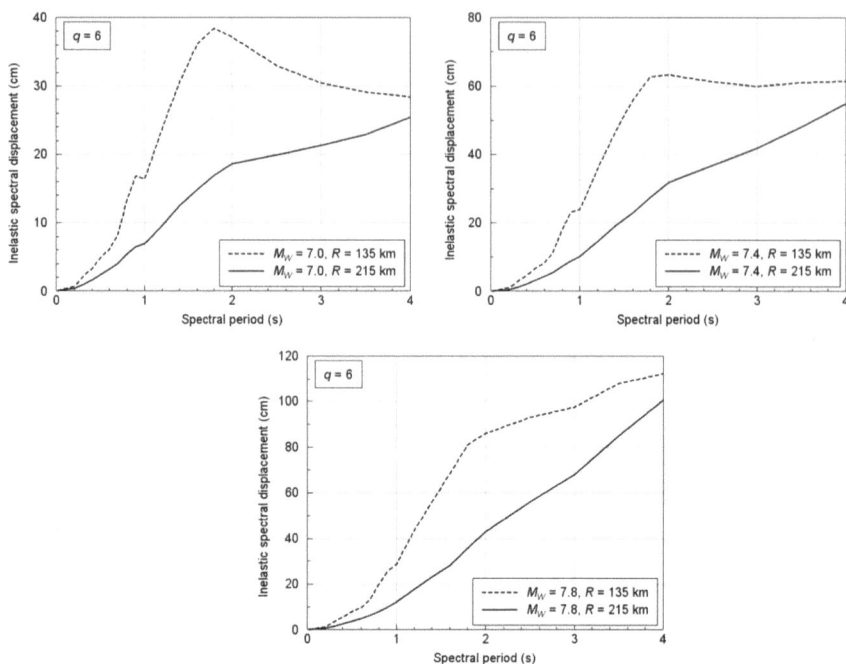

Fig. 5.5 Comparison between the median inelastic spectral displacements function of the earthquake magnitude and source-to-site distance for a behaviour factor $q = 6$

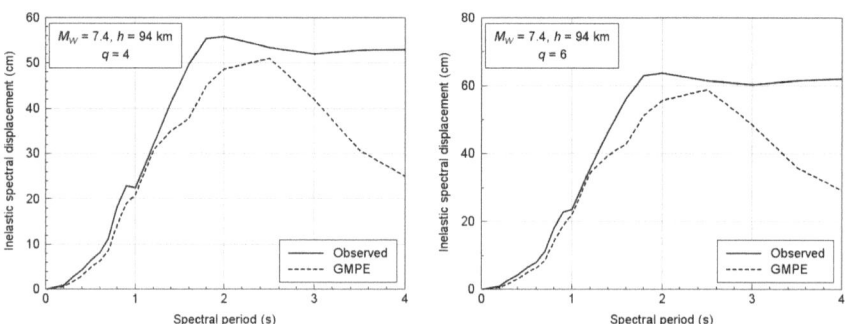

Fig. 5.6 Comparison between the observed and predicted inelastic spectral displacements for INCERC station during the Vrancea 1977 earthquake for $q = 4$ (*left*) and $q = 6$ (*right*)

Table 5.1 Behaviour factors according to P100-1 for concrete buildings	Ductility class	Behaviour factor, q
	DCH	2.00…6.75
	DCM	1.50…4.75
	DCL	1.50…2.00

5.2   Challenges in the Seismic Design

The main challenge in the design of multi-storey reinforced concrete buildings in Bucharest is due to the particular shape of the acceleration design spectrum. In the Bucharest area, the spectrum has a long control period T_c of 1.6 s, placing most mid-rise concrete buildings in the constant acceleration region of the design spectrum. As a rule of thumb, for mid-rise concrete buildings in Bucharest (with 6–15 storeys), the natural period for seismic analysis can be estimated as the number of storeys multiplied by 0.1. The design peak ground acceleration in the Bucharest area, as prescribed by the current Romanian seismic design code P100-1/ 2013 (2013), is 0.3 g. From the designer's point of view, the design peak ground acceleration is not particularly high, but very aggressive spectral acceleration values are prescribed for mid-rise buildings because of the shape of the acceleration spectrum (Fig. 5.7). In the Bucharest area, the value of spectral acceleration is 0.75 g for all elastic systems with natural periods of vibration situated between 0.32 and 1.6 s (control periods T_B and T_C).

As explained above, the spectral accelerations for mid-and high-rise buildings taller than 28 m have to be further amplified with importance/exposure factors of either 1.2 or 1.4. As the increase of building height commonly results in a shift of the natural vibration period, structures with 12–15 stories are the most sensitive because their fundamental vibration lies in the constant acceleration region of the spectrum amplified by the importance factor of 1.4 (Fig. 5.8). Design spectral acceleration (denoted as S_d in Fig. 5.9) for an elastic system with a natural period of vibration of 1.6 s can be as high as 1.05 g, if an importance factor of 1.4 is considered.

Because of these large spectral ordinates of the constant acceleration region of the acceleration design spectrum and the particularly high value of the predominant period, a capacity design procedure for RC buildings with relatively large behaviour factors is prescribed by the code. For coupled shear wall concrete structures taller than 45 m, with natural periods smaller than 1.6 s, the design spectral acceleration

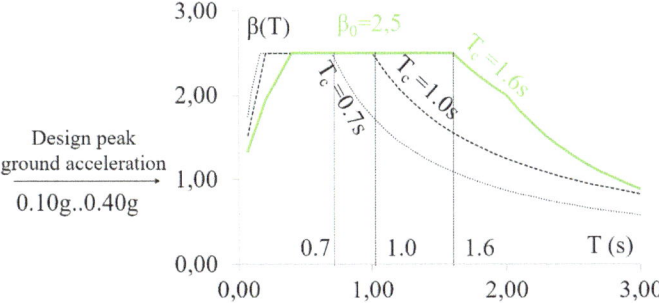

Fig. 5.7 Normalized acceleration design spectra in Romanian seismic design code P100-1/2013 (2013)

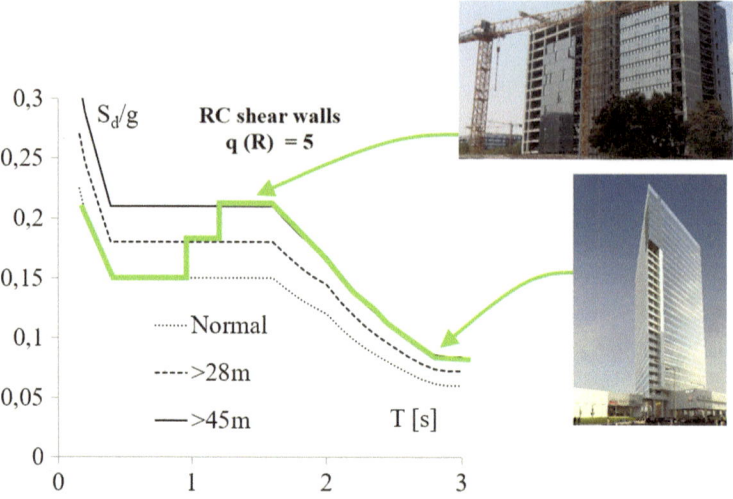

Fig. 5.8 The shape of the acceleration design spectrum for various values of the importance factor

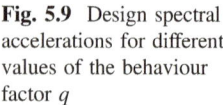

Fig. 5.9 Design spectral accelerations for different values of the behaviour factor q

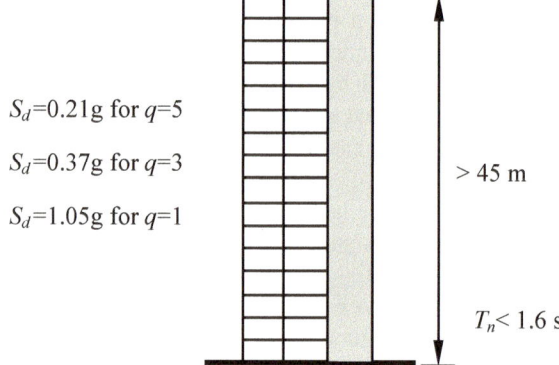

S_d=0.21g for q=5

S_d=0.37g for q=3

S_d=1.05g for q=1

is 0.21 g, if a behaviour factor of 5 is considered. For the same building a behaviour factor $q = 3$ results in spectral acceleration values of 0.37 g. An attempt to design such structures for an elastic response under the design earthquake would increase the spectral acceleration to 1.05 g (Fig. 5.9).

Spectral acceleration values of 0.25 g can hardly be used in the practical design situation for mid-rise buildings (values obtained by dividing the elastic spectral acceleration by the behaviour factor q). Even design spectral accelerations over 0.15 g generate serious issues in structural design. In Bucharest, very heavy mid-rise building structures with thick shear walls, robust frames, stiff and strong basements, piles and foundation mats usually result from the design process. It is now common in Bucharest to have foundation mats up to 2.0 m thick for regular mid-rise buildings with steel consumption in excess of 150 kg/m^3 of concrete.

(a)   (b)

Fig. 5.10 Heavily reinforced shear wall (**a**) and foundation mat (**b**) in a building in Bucharest

Shear walls, 400–800 mm thick, are heavily reinforced as well (Fig. 5.10), the steel consumption exceeding in most cases 150 kg/m^3 of concrete. Podium slabs with a thickness of 300–400 mm are used to transfer the seismic forces from the upper structure to the basement walls. Informative values of the steel consumption for mid-rise buildings in Bucharest are given in Table 5.2. Concrete consumption averages 0.6 m^3/m^2 of the total area of the building. These large consumption indexes raise serious concerns for real-estate developers, most of whom come from western Europe and can barely understand the fundamentals of seismic protection and the peculiarities of the seismic hazard in Bucharest.

Controlling the lateral displacement under design and service earthquakes is a serious issue in structural design in Bucharest. An appealing strategy to reduce the seismic response for mid-rise buildings is to increase the fundamental period of the structure by decreasing the lateral stiffness. If the natural period is longer than the control period $T_C = 1.6$ s, then the spectral acceleration starts to decrease. In this way, slender structures able to accommodate the strength and ductility demand can be designed. However, in many situations it is difficult to control the lateral drift especially under service earthquake loads. Any attempt to control the drift by increasing the lateral stiffness of frames and shear walls would shift the natural period of the structure back towards the constant acceleration region of the design spectrum (Fig. 5.11). Highly engineered structures with tailored stiffness and strength over the height, as presented in Fig. 5.12, are necessary in some situations to allow the control of drift, lateral strength, plastic mechanism and ductility.

Element type	Steel (kg)/concrete (m^3)
Shear walls	150–165
Basement walls	140–150
Beams	180–200
Columns	170–190
Foundation mat	150–165
Slabs	120
Podium slabs	180

Table 5.2 Informative values of steel consumption for mid-rise buildings in Bucharest

Fig. 5.11 Increase in the design spectral acceleration with structural stiffness for medium-rise and high-rise buildings

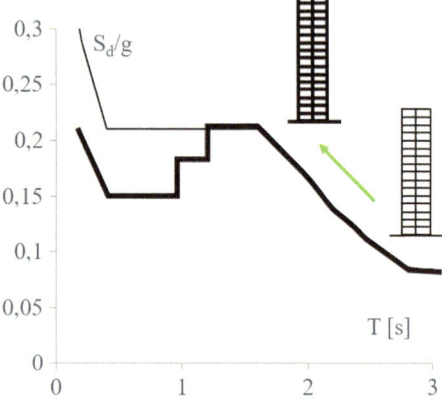

Fig. 5.12 Structural layout in elevation for controlling both lateral displacements and strength

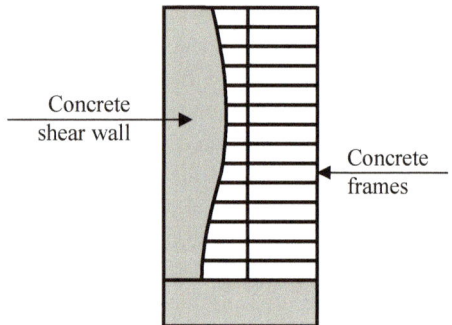

The cross-section of the shear walls in elevation is shown in grey in Fig. 5.12. One can observe that the largest shear wall cross-section is encountered above mid-height. However, the control of the seismic response of such structures requires advanced analysis methods, such as nonlinear static analysis or non-linear time history analysis. Many design offices in Bucharest regularly use such types of advanced analysis methods for both reinforced concrete and steel buildings.

The design seismic action in Bucharest generates top lateral displacement demands for mid- and high-rise buildings in excess of 60 cm. For importance class I buildings, lateral displacement demand goes beyond 80 cm. There is little international experience regarding the behaviour of regular buildings under such large lateral displacements. In dual concrete structures, the rotation of beams exceeds 0.03 rad in many situations and in some situations the rotation of coupling beams exceeds 0.06 rad. The rotation of shear walls in the plastic region usually exceeds 0.01 rad. To achieve the necessary ductility, the reinforcement in plastic regions is connected using mechanical couplers. Given the lateral rotational demand, these couplers are type S2 according to the definition given in ACI 318-11 (2011) which are able to sustain large plastic deformations of rebars without the failure of the connection. There are serious issues regarding the strict control of steel quality, in order not to exceed the maximum ultimate tensile strength, and the control of mechanical couplers certified as type S2, including site control for couplers installation. In addition, there are strong concerns regarding the behaviour of non-structural components under such lateral displacements, repairing costs and downtime.

5.3 Structural Systems for Buildings in Bucharest

In Romania reinforced concrete is the most widely used structural material, and is employed in the construction of residential buildings, office buildings, hospitals, hotels, etc. The first reinforced concrete structures in Romania were erected at the end of the XIXth century under the supervision of the engineer Anghel Saligny. Following the rapid urbanisation that occurred in Romania at the beginning of the 1960s, reinforced concrete was the main material used for multi-storey residential buildings. Typified building projects, many of them based on precast concrete elements, were used in most of the major cities. The revival of the construction industry at the end of the 1990s also relied on concrete as the main structural material. This was mainly due to the cost effectiveness of RC structures in comparison with steel structures in Romania. Concrete quality steadily increased from classes C 8/10 to C 16/20 in the 1960s, to classes C 25/30 to C 40/50 at the present time.

Architectural requirements have changed a great deal in the last 15 years, causing a shift in the structural layout of multi-storey buildings. In the 1960s, in multi-storey residential buildings, honeycomb weakly reinforced shear walls structures were mainly used. The shear wall density was rather high. Nowadays, heavily reinforced structural elements with limited concrete sections are used.

In most European countries the structural design of buildings is governed by gravity or wind action. Simple structures with thin structural elements and low redundancy, strictly adapted to the architectural requirements result as a consequence. Due to the relatively high seismicity of Romanian territory, the structural design of concrete buildings is governed by seismic action. Therefore, structural systems currently used in Romania significantly differ from the European practice. Seismic design in Romania largely follows the international practice of countries with high seismicity, such as the United States or New Zealand.

Residential buildings in suburban areas are low height structures of one or two storeys with concrete frame structures and thick masonry partitions (Fig. 5.13). Rather short openings, ranging from 4 to 5 m, are commonly used. Concrete columns have cross-sections of 30 cm × 30 cm and beam height varies between 40 and 50 cm. 15 cm thick slabs are commonly used. The infilled masonry walls are 25 or 30 cm thick. The concrete quality is rather low, not exceeding class C 20/25 in most cases.

Fig. 5.13 Two storey residential building in Bucharest suburban area

Fig. 5.14 One storey precast concrete structure

In the retail sector, single storey precast structures are mainly used (Fig. 5.14). The main span ranges from 25 to 40 m and the secondary span varies between 7 and 15 m. Story height commonly ranges from 8 to 20 m. Columns are designed as vertical cantilevers embedded in pocketed foundations with hinged prestressed beams resting at the top. In the Bucharest area, column sizes are rather large, ranging from 80 cm × 80 cm to 100 cm × 100 cm. The concrete quality for precast elements exceeds class C 40/50.

For multi-storey residential buildings with more than three or four storeys, shear walls structures are mainly used (Fig. 5.15). Relatively short beam openings

Fig. 5.15 Multi-storey residential building in Bucharest

ranging from 4 to 6 m are used. Concrete floors consisting of slabs supported by beams are generally used. Due to architectural requirements the use of flat slabs is becoming increasingly popular. Vertical structural elements consist of concrete walls or columns embedded in a dense network of partition walls. The structural layout is often irregular, but redundancy and overstrength requirements are met. Partition walls are built using mainly large hollow clay masonry units. The concrete quality varies in general from class C 20/25 to class C 30/37.

In most office buildings, the lateral load resisting structural system consists mainly of a coupled shear walls concrete core located in the central area of the building (Figs. 5.16 and 5.17). For improved stiffness and redundancy, inner and outer concrete frames are added. If the architectural requirements prohibit the installation of interior beams, only outer frames are used. The use of a central concrete core with flat slabs is generally prohibited by structural requirements as prescribed by the code. However, there is steady pressure from the industry for this structural system to be adopted on a large scale.

Urban regulations in Bucharest require office buildings to have a certain number of parking spaces depending on the total number of residents. The parking areas are located in the basements and basically dictate the main openings of the upper structure. Eight meter openings are highly popular among architects because of the possibility of accommodating three parking spaces in between two adjacent columns. In this segment of multi-storey buildings, higher concrete classes, ranging from C 30/37 to C 40/50, are generally used.

The thickness of the shear walls usually results from the shear strength requirement. According to P100-1/2013 (2013), the allowable average shear stress in the plastic region of a shear wall designed for DCH is 0.15 f_{cd} where f_{cd} is the

Fig. 5.16 Multi-storey office building in Bucharest

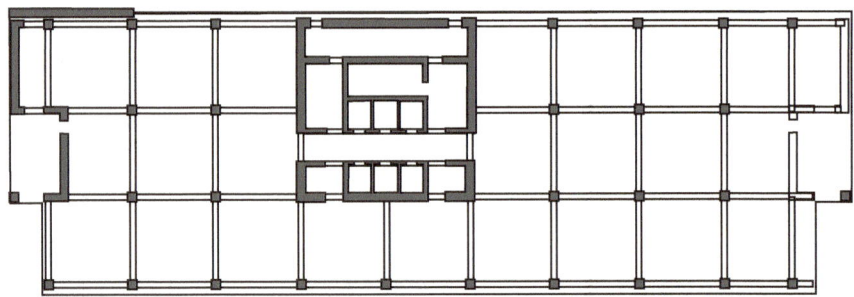

Fig. 5.17 Typical floor layout for an office building in Bucharest

design compressive strength of concrete as defined in EN 1992-1-1 (2004a). This means that the allowable average shear stress ranges from 3 MPa for C 30/37 to 4 MPa for C 40/50.

As a rule of thumb, 1.0 m² of the shear area of concrete walls is necessary for each 1000 m² of building area, in each principal direction of the building. Concrete shear wall density calculated as the ratio between total area of shear walls and total gross area of the building should be larger than 0.2%. This value is estimated based on a multi-storey building located in Bucharest with the natural period lower than 1.6 s considering the following parameters: behaviour factor of $q = 5$, importance factor of 1.4, base shear coefficient of 0.22, concrete class C 40/50, flexural overstrength factor of 1.5, unit weight of the building 12 kN/m². The shear wall density for the Bucharest area is two times larger than that of the Chilean practice (Lagos 2017). Despite moderate values of design peak ground acceleration (0.3 g), the shape of the design acceleration spectrum generates high base shear coefficients. In most cases, shear wall thickness determined by design ranges between 400 and 800 mm, but higher values up to 1200 mm can be found in practice, as well.

According to the Romanian building standards, transversal reinforcement in the plastic region of a shear wall should be able to transfer all the shear force considering an angle of the compression strut of 45°. Basically, transversal reinforcement in the plastic region of shear walls should be able to suspend the entire seismic design load multiplied with the flexural overstrength factor. If the concrete section is fully used to transfer an average shear stress of 4 MPa (concrete class C 40/50 as explained above), a necessary shear reinforcement ratio of 0.009 can be calculated for S500 steel (f_{yd} = 435 MPa). This ratio is rather high, generating the need for large diameter rebars and several curtains of reinforcement. For example, for a 600-mm thick wall it is necessary to install 5400 mm^2 of transversal reinforcement. Three curtains of reinforcement with 20 mm diameter rebars spaced at 175 mm are therefore necessary.

In mid-rise buildings in the Bucharest area, the beam and columns sections usually result from lateral drift limitation criteria for serviceability requirements. This is obvious for concrete moment resisting frames, but it is also true for dual concrete structures, as well. Because of the large lateral displacement demand, concrete shear walls alone cannot restrain the lateral deformation of the building at the upper stories. The shear walls exhibit a cantilever deformation pattern with the largest rotation at the top stories if the surrounding frames are weak, being designed only for gravity loads. This is likely to cause severe distortion of the panels located between two adjacent walls and heavy damage to the nonstructural elements (Fig. 5.18).

To restrain the lateral deformation of the walls in the upper stories it is necessary to connect the shear walls with strong frames or to increase the coupling of the walls. The height of the outer beams cross-section is usually larger than 1/10 of the clear opening of the beam. In the inner spans, the height of the beams' cross-section is strictly restrained by architectural requirements to 1/14–1/12 of the clear opening of the beam. The outer frames can transfer a significant part of the total overturning moment by generating high axial forces in the corner columns.

Fig. 5.18 Deformation pattern for a concrete shear wall building with weak frames

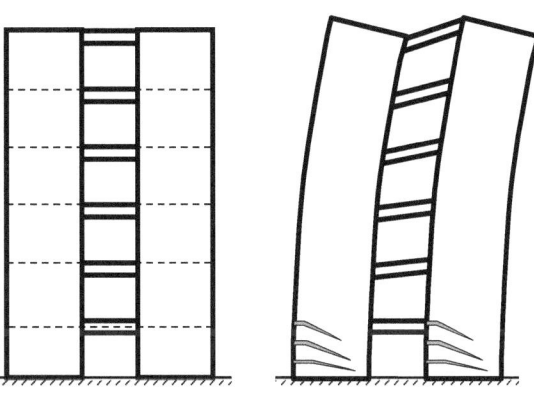

In dual structural systems, the columns are subjected mainly to axial loads. Axial loads from gravity loads are highly amplified by seismic action in the corner columns. The cross-sections of the columns are commonly determined by ductility related requirements regarding limitation of axial force ratio in the seismic design situation. The allowable maximum value of axial force ratio calculated using the design compressive strength of concrete is 0.45 if the columns are part of the structural system resisting to seismic action. In dual systems, the columns' cross-sections can be smaller because of the interaction with the shear walls that limit the rotational ductility demand in columns in the ground floor. However, in addition to the rotational ductility of the columns, there are other issues that need to be addressed when deciding the columns' cross-sections: anchorage of the beams' longitudinal reinforcement in beam-column joints and shear strength of beam-column joints.

According to P100-1/2013 (2013), the anchorage length of the longitudinal reinforcement in beams should be calculated in accordance with the provisions of EN 1992-1-1 (2004a), considering a maximum stress in the longitudinal reinforcement in critical regions of beams of 1.2 f_{yd}. Total anchorage lengths ranging between 49 φ and 82 φ result for concrete classes between C 40/50 and C 20/25, where φ is the diameter of the bar being anchored (Table 5.3). Compared to similar standards, EN 1992-1-1 (2004a) appears to be highly conservative with respect to anchorage lengths especially for compressed rebars located at the top of the beams' cross-sections (Cotofana et al. 2017). According EN 1992-1-1 (2004a), anchorage length of compressed rebars is larger than anchorage lengths of rebars under tension.

To accommodate such long anchorage lengths, it is necessary to increase the width of the columns. As a rule of thumb, the width of the column cross-section measured in the direction of the beam should not be smaller than 25 φ, where φ is the maximum diameter of the rebar being anchored in the joint. It is emphasized here that, even for dual structural systems, beams in upper stories are expected to fully yield under the design ground motion and proper anchorage of the reinforcement at full strength of the steel is strictly required. The relation between the minimum column width, beam height and mandrel diameter from anchorage requirements for any diameter φ bottom beam rebars, based on the prescriptions of P100-1/2013 (2013) and EN 1992-1-1 (2004a), is presented in Fig. 5.19. These

Table 5.3 Anchorage lengths for beam rebars	Concrete class	Anchorage lengths, l_{bd}	
		Bottom rebars	Top rebars
	C 20/25	57φ	82φ
	C 25/30	48φ	69φ
	C 30/37	43φ	62φ
	C 35/45	39φ	56φ
	C 40/50	34φ	49φ

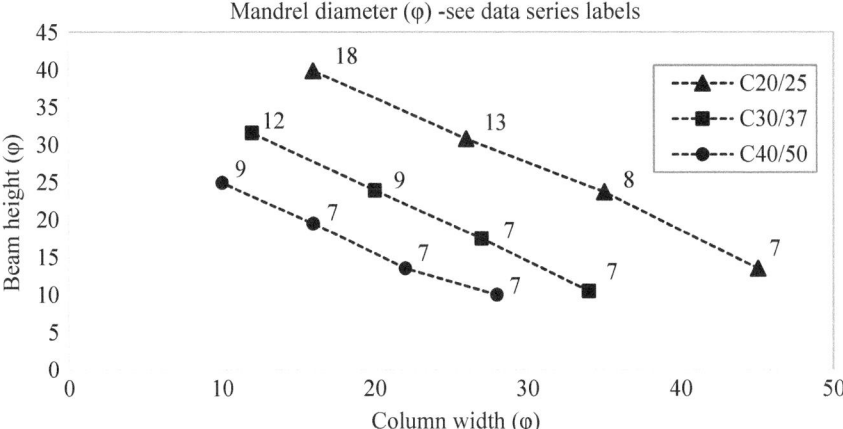

Fig. 5.19 Relation between the minimum column width, beam height and necessary mandrel diameter from anchorage conditions for beams' bottom rebars (diameter φ)

values are higher for top rebars because the allowable steel-concrete bonding stress is lower.

Considering these arguments, buildings with dual structural systems in Bucharest have columns with square or circular cross-sections ranging from 600 to 1000 mm width and longitudinal reinforcement ratio ranges between 1 and 2%. The minimum longitudinal reinforcement ratio in columns prescribed by P100-1/2013 (2013) is 1% for DCH, the same as in EN 1998-1 (2004b).

As the design effort is oriented towards high ductility structures, special attention is paid to the detailing of the structural components. Spacing of the transversal reinforcement in columns and beams commonly results from ductility related criteria concerning concrete confinement and the restraining buckling of compressed main rebars. The usual spacing of the shear reinforcement in critical regions of beams and columns is 100 mm, each longitudinal rebar being restrained by a leg of a hoop or tie.

According to Romanian seismic design code provisions, both edges of a shear wall should be additionally strengthened by vertical and horizontal reinforcement. This portion of the shear wall is called the boundary element. The length of each boundary element is one-tenth of the shear wall length. The minimum vertical reinforcement ratio in the boundary elements is 0.5%. Enough transversal reinforcement in each boundary element should be provided to properly confine the compressed area, to prevent buckling of the vertical reinforcement and to allow the full anchorage of the web horizontal reinforcement in the boundary element.

Vertical rebars in shear walls are always placed in the area enclosed by transversal reinforcement. In this way, the buckling of the vertical reinforcement after severe yielding in tension can be prevented. In boundary elements, the maximum spacing of hoop or tie legs is 200 mm. The vertical spacing transversal reinforcement in boundary elements is usually between 100 and 200 mm. A typical

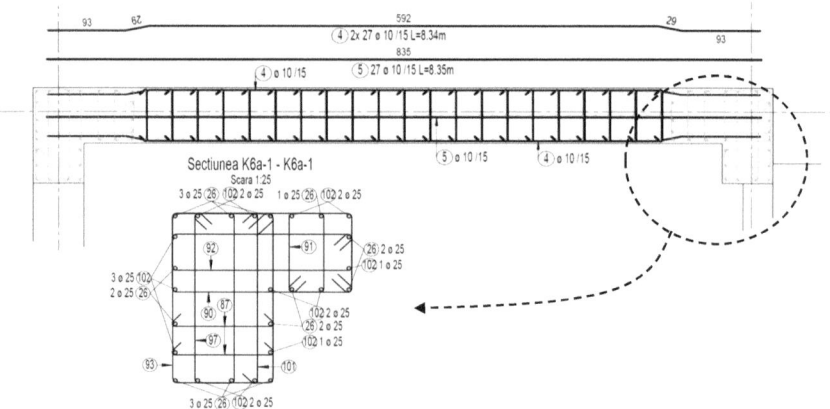

Fig. 5.20 Reinforcement details for a shear wall

shear wall reinforcement detail for modern concrete buildings in Bucharest is given in Fig. 5.20.

In past Romanian building practice, web horizontal reinforcement of shear walls was overlapped with boundary elements stirrups outside of the confined concrete core (Fig. 5.21a). This was fairly acceptable for small diameter horizontal rebars (φ 8 mm, φ 10 mm) used in traditional RC buildings. Given the high lateral seismic displacement demand in the Bucharest area currently prescribed by the code, relative high rotation of shear walls at the base can be generated and a certain degree of compressed concrete crushing should be expected. This would alter the overlapping connection of horizontal reinforcement and reduce the effectiveness of transversal reinforcement. Therefore, to accommodate the high ductility demand, in the case of large diameter rebars horizontal web reinforcement should be properly anchored within the confined concrete core (Fig. 5.21b).

Fig. 5.21 Solution for the reinforcement of the boundary elements

(a)

(b)

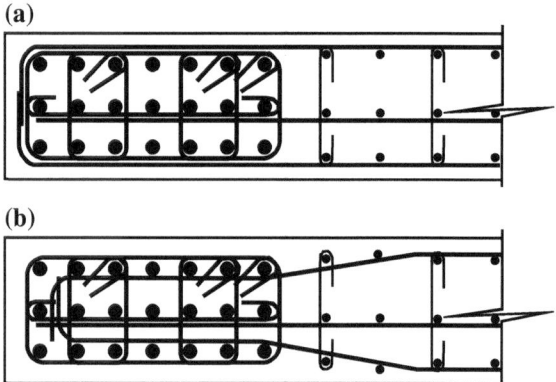

The relatively large values of the base shear coefficients for medium-rise and high-rise buildings in Bucharest result in difficulties in connecting shear walls with horizontal diaphragms, especially at the transfer (podium) slab. According to P100-1/2013 (2013), three shear transfer mechanisms can be considered: shear friction along the length of the vertical element, direct bearing compression forces at wall ends, and tensile forces through collector reinforcement. The capacity of shear friction is limited by the amount of connection reinforcement perpendicular to the web of the shear wall and the tensile strength of concrete. Direct bearing compression is limited by the compressive strength of concrete and tensile forces in the collector reinforcement are limited by the amount of reinforcement that can be placed at the intersection between the shear wall and the slab. If this amount is not properly considered, wall shear sliding failure mechanism can occur because of poor continuity of concrete at the interface. The central concrete core is, in many cases, isolated from the podium diaphragm by large MEP (mechanical, electrical, plumbing) openings, making the transfer of the shear forces even more difficult. In such a situation, the role of the collector reinforcement increases as the shear-friction cannot be fully considered.

5.4 Protection of Non-structural Components

One of the main concerns related to seismic design for large lateral displacements is the protection of non-structural components. Many non-structural components have a brittle behaviour and cannot accommodate large deformations. The most sensitive structural components are the masonry partition walls. Partition walls made of masonry are very popular throughout Europe, from Portugal to Turkey. This popularity is generated mostly by tradition, cost, locally available bricks and mortar and readily available skilled labour. Advances in the manufacturing industry favoured the development of hollow clay masonry units with very thin interior walls. These units have excellent thermal insulation properties, their large size and low weight favouring the speed of the construction process. In seismic areas the main disadvantages of these bricks are their brittle behaviour and limited repair options.

Tests conducted at the Technical University of Civil Engineering of Bucharest showed that the in-plane behaviour of masonry walls built using large hollow clay masonry units is rather brittle. The first diagonal shear crack appears at approximately 0.5% lateral drift and a complete loss of the bearing capacity occurs at 1–1.5% lateral drift angle (Fig. 5.22).

The vulnerability of non-structural masonry walls was observed by the authors during a post-earthquake investigation mission in the city of Van, Turkey, carried out within the international UNESCO IPRED Platform (http://www.unesco.org/new/en/natural-sciences/special-themes/disaster-risk-reduction/geohazard-risk-reduction/networking/ipred/). The city of Van was strongly affected by two earthquakes in the fall of 2011: Van-Ercis earthquake on October 23 ($M_W = 7.1$) and

Fig. 5.22 Damaged masonry panel at 1% lateral drift as recorded in structural testing at the Technical University of Civil Engineering, Bucharest

Van-Edremit earthquake on November 9 (M_W = 5.6) (EERI 2012). Masonry infills were built, according to the construction practice in the region, using hollow clay units. The width of the masonry panels was 30 cm for exterior walls and 15 cm for interior walls.

Despite the minor structural damage consisting mainly of shear and flexural cracks in the critical region of beams, masonry panels were heavily damaged in many of the inspected buildings (Popa et al. 2013). Most of the interior and exterior masonry panels on the first three stories in multi-storey buildings sustained severe damage that was beyond repair (Fig. 5.23). The replacement of these panels is obviously necessary for the future use of the building. Full or partial collapse of some masonry panels was n also observed. The overturning of gable walls or double leaf outer masonry panels was also visible.

Structural analysis performed on one of the inspected buildings revealed sufficient strength according to the provisions of the Turkish earthquake standard at the time of construction (Popa et al. 2015). Lateral overstrength factors of around 2.0 were determined. After the earthquake, the design acceleration prescribed by the Turkish code was increased by 25%, but the lateral strength of the buildings was still sufficient. The lateral displacement demand, obtained using the capacity

Fig. 5.23 Non-structural damage after the Van earthquake in Turkey (2011)

Fig. 5.24 Non-structural masonry details in Romania

spectrum method or linear time-history analysis, was relatively low with peak lateral drifts of around 0.5%. This includes the amplification caused by the torsional response observed using linear time-history analysis. Nonlinear time-history analysis, performed based on loading scenarios comprised of two successive earthquakes, revealed larger values of the displacements. A maximum 1% lateral drift was calculated based on the most severe loading scenario.

Similar observations related to the sensitivity of non-structural masonry elements in modern concrete buildings were made in Italy after the Abruzzo earthquake of 2009 (D'Ayala and Paganoni 2009) and after the Emilia earthquake of 2012 (Lagomarsino 2012).

Non-structural masonry construction practice in Romania is similar to the existing practice in Portugal, Italy or Turkey (Fig. 5.24). Extensive non-structural damage can occur after the design earthquake due to high displacement demands imposed by Vrancea intermediate-depth earthquakes. The use of masonry partitions should be avoided whenever possible. These partitions should be replaced by engineered non-structural walls with high deformation capacity. Fortunately, in most office buildings masonry partitions are no longer used.

Fig. 5.25 Gap between masonry partition and outer frame

Detaching the partitions from the main structural elements is a viable option, but care should be taken to prevent overturning. This might require the construction of a light frame, surrounding the partition, detached from the main structural frame by a 5 cm gap (Fig. 5.25). The interaction of partition walls with the main frame columns or the plastic regions of the beams should be prevented. Care should be taken as detaching the masonry from the surrounding frame with a clear gap raises fire safety concerns and, in many situations, is not approved by the fire department.

5.5 Structural Design of an Office Building in Bucharest—A Brief Example

In this section, some results of a preliminary design procedure for dual structural systems are presented. This preliminary design procedure is based on structural equilibrium after full yielding of the structure under lateral seismic loads.

The building in Fig. 5.26 has 6 longitudinal spans of 8.10 m and 4 transversal spans of 8.10 m. There are 12 storeys above ground and three underground levels. The storey area is 1615 m^2 and the total upper story area is 19,384 m^2. The story height h_s is 4.25 m in all stories. The total height above ground is $H = 51$ m. Because of this height, according to the Romanian code, the building is classified as having importance class I. The structural system consists of a concrete central shear wall core, outer frames and flat slabs. In the underground levels outer concrete walls are added.

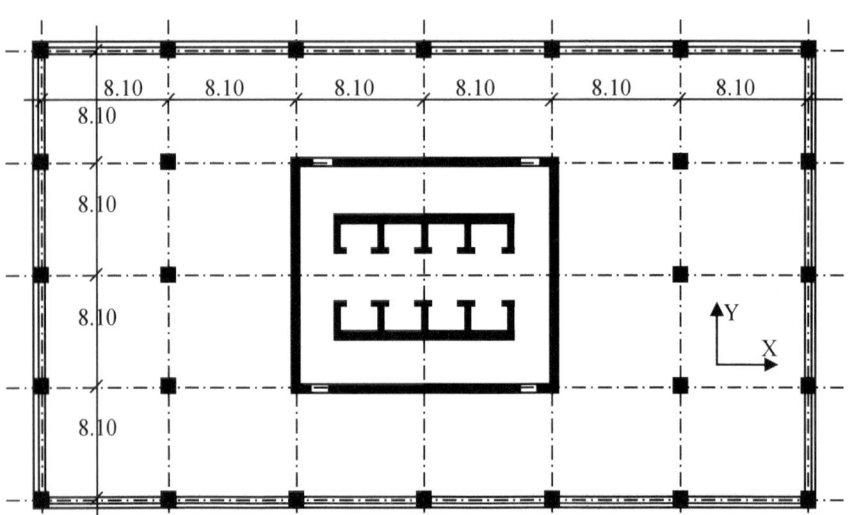

Fig. 5.26 Structural layout for the analysed building

The site is characterized by a design peak ground acceleration $a_g = 0.3$ g, a spectral amplification factor for the constant acceleration region of the design spectrum $\beta_0 = 2.5$ and a control period of acceleration design spectrum of $T_c = 1.6$ s.

The building is designed for the ductility class DCH, considering a behaviour factor $q = 5$. Considering a first natural period of the building of $T_1 = 1.2$ s $< T_c = 1.6$ s, the base shear coefficient can be calculated as follows:

$$c = \gamma_{I,e} \frac{\beta(T)a_g}{q} \lambda = 1.40 \frac{2.5 \times 0.3\,\text{g}}{5} \times 0.85 \approx 0.18\,\text{g}$$

Unit weight of the building is 14 kN/m² resulting in a total weight of 271 MN. The design seismic load is:

$$F_b = c \cdot m = 0.18\,\text{g} \times 27.1 = 48.8\,\text{MN}$$

Total building overturning moment above the podium floor generated by F_b can be approximated by (Fig. 5.27):

$$M_{overturning} = 2/3HF_b = 2/3 \times 51 \times 48.8 = 1660\,\text{MNm}$$

5.5.1 Outer Frames

The overturning moment is distributed in unequal shares to the outer frames and to the concrete shear wall core. The beams contribute to frames mainly with their

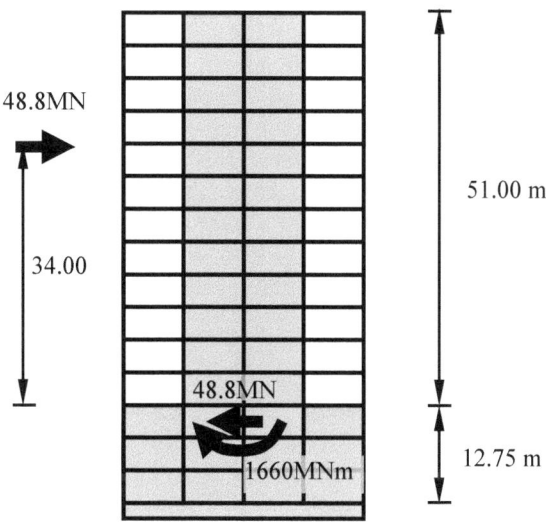

Fig. 5.27 Overturning moment and base shear force for the analysed building

48.8MN

34.00

48.8MN

1660MNm

51.00 m

12.75 m

flexural strength and stiffness. The columns carry the axial load both from the gravitational loads and from the seismic action, and their flexural contribution is negligible in dual systems. The amount of overturning moment that is to be transferred by the frames is limited by the flexural and shear capacity of beams and axial loads in the corner columns. The outer frames have an important contribution to the stiffness of dual structural systems as they are able to limit the drift in the upper part of structure.

In order to limit the overturning moment acting on the central core, it is desirable to transfer it as much as possible through the frames bearing in mind the following limitations:

– To preserve a ductile flexural failure mode for the outer beams;
– To maintain enough bending moment in the central concrete core so as to develop a ductile flexural failure mode;
– To limit the axial load ratio in the concrete columns so as to maintain a certain ability to exhibit flexural plastic deformations in the ground floor.

In this respect, apart from architectural constraints, the height of beams should be large enough for suitable strength and stiffness. The height of the beams' cross-section is usually larger than 1/10 of the clear span. For a story height of 4.25 m and clear spans of 7.40 m, outer beams with cross-sections of 400 mm 850 mm can be used in the longitudinal direction (X direction). In the transversal direction (Y direction), as the lever arm of the axial forces in the corner columns caused by seismic load is smaller, increased section of the beams is advisable. In this example, transversal beams of 400 mm × 1000 mm are used.

The width of the beams should be large enough to accommodate the longitudinal reinforcement on one or two rows at most. For a proper rotational ductility, the width of the beams should be at least 1/3 of their height.

Despite the presence of the shear walls, the columns in the corners are expected to bend over their yield limit in the ground floor, with drift angles exceeding 1%. Therefore, it is wise to limit the axial load ratio in these columns to a level that can prevent any type of brittle failure. The axial load ratio, v, calculated based on the design compressive strength of concrete, f_{cd}, is commonly limited to 0.55 in such columns. The corresponding value of the axial load ratio for the medium compressive strength of concrete is 0.30. The maximum axial load for a 800-mm square column can be calculated as:

$$N_{Ed} = v b_c h_c f_{cd} = 0.55 \times 800 \times 800 \times 40/1.5 = 9386\,\text{kN}$$

Considering an axial force from gravity loads of approximately 3327 kN, calculated based on a unit weight of 14 kN/m^2, approximately 6000 kN axial force in the most compressed corner columns can be further used to transfer the building overturning moment.

Considering that most beams yield under the design earthquake, the corresponding shear force in the beams in the Y direction (Fig. 5.28) is nearly 550 kN (approximated here as 6000/11). The corresponding bending moment is 1990 kNm.

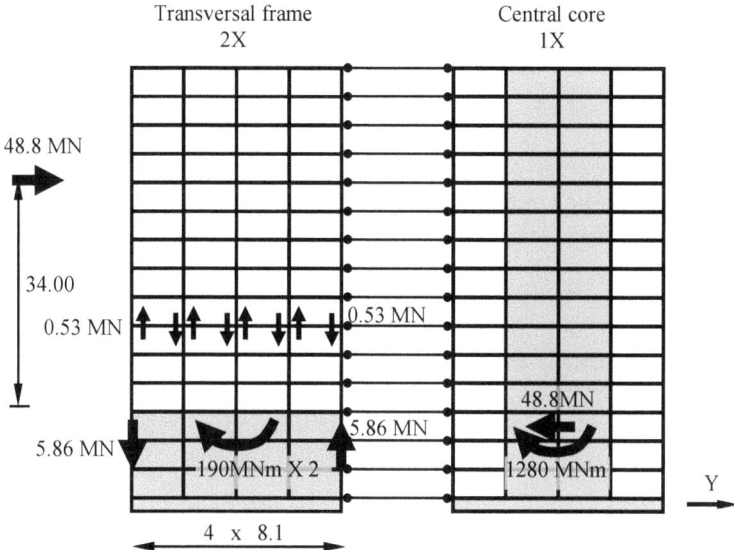

Fig. 5.28 Simplified distribution of overturning moment and shear at full plastic mechanism (Y direction)

Beams can be reinforced with 11 φ 25 mm rebars at each face generating a moment capacity of 1950 kNm and a corresponding shear force of 533 kN. At full yield of all 11 beams, the axial force in the corner columns increases with an indirect axial force N_{ind} = 5863 kN from 3300 to 9190 kN. This is equivalent to an axial force ratio of 0.54.

The corresponding overturning moment transferred by both frames aligned with Y direction is:

$$M_{frames}^{Y\,direction} = 2 \cdot N_{ind} \cdot L_y = 2 \times 5863 \times 4 \times 8.1 = 380\,\mathrm{MNm}$$

where L_y = 4 × 8.10 is the lever arm of the axial force in the corner columns measured in the Y direction.

Shear in beams is much lower than the shear capacity associated with failure of concrete in the compressed struts, $V_{Rd,max}$, calculated according to EN 1992-1-1 (2004a), considering a strut angle of 45°. This implies that the beams can transfer the shear associated with their flexural strength, if adequate transversal reinforcement is provided.

In the X direction, the distance between corner columns is larger. Beams with cross-sections of 400 mm × 850 mm reinforced with 9 φ 25 mm can sustain a bending moment of 1496 kNm and a corresponding shear of 410 kN.

The corresponding overturning moment transferred by both frames aligned with the X direction is:

$$M_{frames}^{Y\ direction} = 2 \cdot N_{ind} \cdot L_y = 2 \times 410 \times 11 \times 6 \times 8.10 = 438\,\text{MNm}$$

By loading the corner columns to the axial load ratio limit, out of the total overturning moment generated by the design seismic load of 1660, 380 MNm can be transferred by the frames aligned with Y direction (Fig. 5.28) and 438 MNm in the X direction. Roughly 25% of the overturning moment can be transferred by frames. The cross-section of the corner columns should be further increased to allow for larger axial loads, but still keeping the axial force ratio limit. The rest of the overturning moment and virtually all of the shear force should be transferred by the concrete shear walls core.

This high amount of overturning moment transferred by frames can be observed only by non-linear static analysis. Conventional elastic analysis underestimates the contribution of frames in dual systems. Nonlinear static analysis should be used in design in order to take the full benefit of outer beams' flexural strength and reduce the bending moments in the central core and in the foundation mat.

In order to achieve the necessary ductility for columns, the Romanian seismic design code P100-1/2013 (2013) prescribes a minimum longitudinal reinforcement ratio of 0.01. For a 800-mm square column 8 φ 25 mm + 8 φ 20 mm longitudinal rebars would be required. In Romanian practice, each longitudinal rebar is connected with a stirrup or tie leg. Usually, spacing of stirrups is 100 mm in critical zones and 200 mm elsewhere.

5.5.2 Shear Walls

If in-plane flexural response of shear walls with large lateral deformations is expected, the web thickness of the shear walls is usually determined by shear strength requirements. In this kind of dual system, the entire shear force caused by the design seismic load is transferred by the shear walls. The contribution of the columns in this respect can be neglected.

The shear force should be estimated based on the design horizontal seismic load, flexural overstrength and contribution of higher modes.

The overstrength factor should account both for the flexural overstrength of the outer beams and for the overall flexural overstrength of the concrete shear walls core. In most practical design situations, for such structures beams and walls can be suitably reinforced in order not to generate an additional increase of the overturning moment. In this example, it is difficult to increase the contribution of the frames due to architectural limitations in both beams and columns. Therefore, the frames can be designed so as to exhibit no overstrength aside from the overstrength generated by the strain hardening of steel.

A higher flexural overstrength of the frames and of the shear walls by increasing the amount of flexural reinforcement might look appealing for designers, but it should be remembered that if the flexural capacity is increased, the sensitivity of the structure to brittle failure modes is increased. Considering the characteristics of strong ground motions in Bucharest, it is obvious that such structures will severely yield and sufficient ductility should be provided by the design.

The allowable average shear stress in the plastic regions of a shear wall designed for Ductility Class High (DCH) is $0.15 \cdot f_{cd}$, where f_{cd} is the design compression strength of concrete. f_{cd} is calculated by dividing the characteristic compression strength, f_{ck}, to 1.5.

If all the shear force generated by the horizontal design seismic load is to be transferred by the concrete core, the total shear walls web areas aligned with one principal direction can be determined as follows:

$$\sum A_w = \frac{1.2 \times 1.25 \cdot F_b}{0.15 \cdot f_{cd}} = \frac{1.2 \times 1.25 \times 48.8 \times 10^6}{0.15 \times 26.7} = 18.3 \, \text{m}^2$$

where 1.2 accounts for the influence of higher vibration modes over the shear force distribution and 1.25 accounts for flexural overstrength caused by post-yielding strain hardening.

This area of shear walls should be equally placed in both the X and Y direction. In this example, in the X direction there are basically four large shear walls, each 9 m long. In the Y direction, there are two large 16.8 m long shear walls. The width of all these walls should be 600 mm, in order to result in the necessary shear area of 18 sm^2. The cross-section of the shear walls should be large enough to accommodate the necessary flexural reinforcement. The walls dividing the elevator shafts are omitted in this example because those are usually very thin (200 mm) and their capacity is limited.

In this example, the concrete shear walls core should transfer 1280 MNm overturning moment generated by the design seismic load.

When seismic force is acting in transversal Y direction, coupling of the walls YW1 and YW2 with the longitudinal walls XW1 and XW2 is likely to generate noticeable axial loads in XW1 and XW2 (see Fig. 5.29 for labels). This coupling effect may greatly contribute to the overturning capacity of the concrete core. If the shear force in each coupling beam is 750 kN (average shear stress of 1.8 MPa for a 400 mm × 1000 mm rectangular cross-section beam) then the axial force from seismic action in XW1 and XW2 can be approximated as ±18 MN. This is fairly close to the axial load generated by gravity loads in the seismic design situation, so these walls are not expected to be subjected to full tension. This coupling effect would allow the transfer of an overturning moment of 291 MNm.

Walls ES1 and ES2 are neglected in this example when judging the transfer of overturning moment in Y direction because of the short cross-section height in this direction. Therefore, roughly 495 MNm should be transferred by each transversal wall (YW1 or YW2).

Fig. 5.29 Details of the
concrete central core

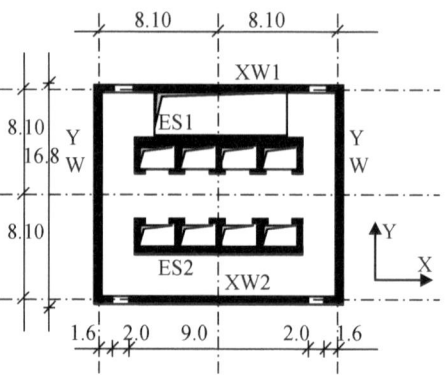

   The axial force contribution to the bending moment capacity is very important,
but in the case of concrete core shear walls the amount of axial force is rather
limited because of the high density of the walls in the central area of the building.
YW1 collects gravity loads from a tributary area of approximately 6.0 m × 24.3 m.
Considering a tributary area of 146 m^2 at each floor and a unit weight of the
building of 14 kN/m^2, the axial force from gravity loads can be approximated as
24,528 kN. The total area of the wall cross-section is approximately 10.83 m^2, thus
resulting in an axial force ratio of 0.08 calculated considering the design compression strength of concrete. This means that the wall is likely to have good ductile
behaviour. The bending moment capacity benefits from a long internal lever arm as
the compression area of the wall at full bending capacity is restrained within the
flange area (Fig. 5.30). The internal lever arm of axial force can be reasonably
approximated as 8.1 m. The contribution of the axial force to the total bending
capacity is roughly 198 MN (24.5 MN × 8.1 m). This is merely 40% of the total
overturning moment to be transferred by this wall (1280/2 MN); the rest of the
296 MN should be attained by vertical flexural reinforcement. Just as an example,
this might require a considerable amount of 50 φ 28 mm vertical rebars in the
boundary elements and 3 φ 20/200 vertical reinforcement in the web.
   According to the currently enforced Romanian seismic design code P100-1/2013
(2013), enough horizontal reinforcement should be provided in the plastic region of

Fig. 5.30 Compressed area
of YW1 and YW2

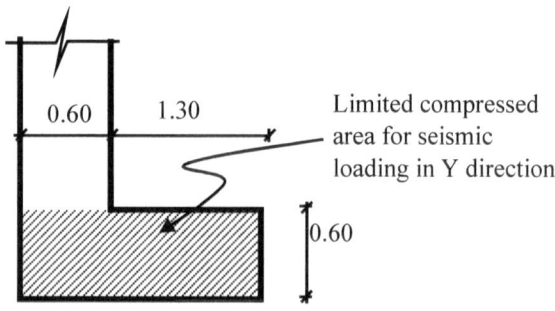

the concrete shear walls to be able to suspend the entire design shear. An inclination angle of the critical shear crack in the plastic region of the shear walls of 45° is prescribed. In flexural controlled walls, the design shear force is the shear corresponding to full flexural yielding of the walls. In this example, there are basically two shear walls in transversal direction Y that should be able to transfer the entire design shear of 1.2 × 1.25 × 48.8 MN. The allowable stress in the reinforcement is the design yielding strength. The necessary horizontal reinforcement area for each transversal wall can be calculated as:

$$\sum A_w = \frac{V_{Ed}}{f_{yd}} = \frac{1.2 \times 1.25 \times F_b}{2 \cdot f_{yd}} = \frac{1.2 \times 1.25 \times 48.8 \times 10^6}{2 \times 435} = 84.2 \times 10^3 \text{ mm}^2$$

If a vertical projection of the critical shear crack of 16.8 m is considered, then 84.2×10^3 mm^2 of horizontal reinforcement can be evenly distributed in 3 curtains of reinforcement with 20 mm diameter rebars spaced at 200 mm. This would total 80.1×10^3 mm^2 of horizontal reinforcement. The reinforcement in horizontal ties located at the shear walls-floors intersection can be supplementarily considered in order to obtain the required area of 84.2×10^3 mm^2. In this example, in order to provide the full anchorage of the web transversal reinforcement (φ 20 mm/ 200 mm), at least 14 mm diameter stirrups spaced at 100 mm should be provided.

5.5.3   Podium Slab

The shear force which is to be transferred at the connection between a shear wall and the podium slab is dependent on the modelling strategy for the podium slab. It is difficult to use accurate estimates of the elastic stiffness for each structural component in the structural analysis. In the case of the transversal shear walls (YW1 and YW), if the podium slab is considered in the analysis as infinitely rigid, the largest transfer force of nearly 40 MN occurs and the bending moments at the base of the wall are smaller (Fig. 5.31). If the slab can deform under in-plane forces, the transfer force decreases to 24 MN, but the bending moments at the base of the wall increase (Fig. 5.32). This results in higher bending moments and shear forces in the foundation mat. This transfer force is estimated when considering elastic flexural and shear stiffness properties of the concrete slab equal to one-half of the corresponding stiffness of the uncracked element. While the infinitely rigid podium slab assumption can be easily disregarded, the purpose of this exercise is to show the range of high values of transfer forces that can occur in mid-rise concrete buildings in Bucharest.

These forces are further amplified with the flexural overstrength factors leading to huge values of nearly 57 MN for an infinitely rigid podium slab and 35 MN for a deformable podium slab. Subsequently, a preliminary design of the podium

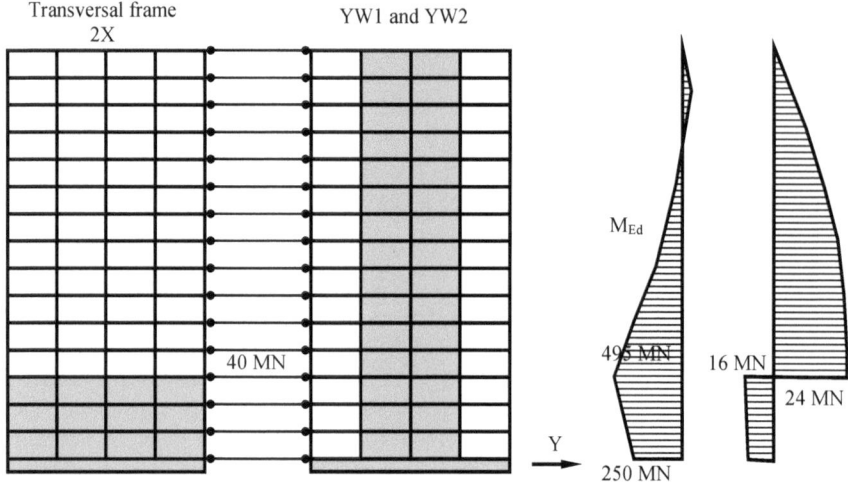

Fig. 5.31 The bending moment and the shear force distribution in transversal walls for in-plane infinitely rigid podium slab

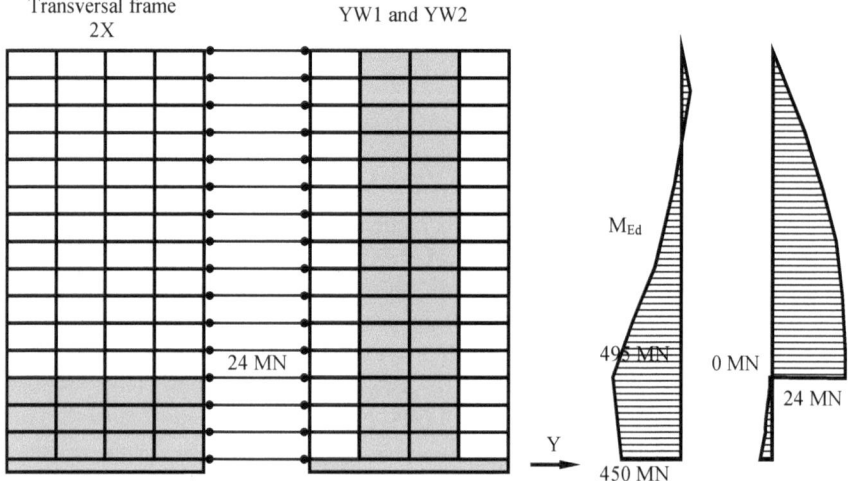

Fig. 5.32 The bending moment and the shear force distribution in transversal walls for in-plane flexible podium slab

diaphragm is presented under the common assumption that the slab is deformable under in-plane forces.

The shear force can be transferred from the wall to the podium slab by three mechanisms: shear friction, direct (compression) bearing and collector (tie) reinforcement (Fig. 5.33). According to P100-1/2013 (2013) and considering a podium

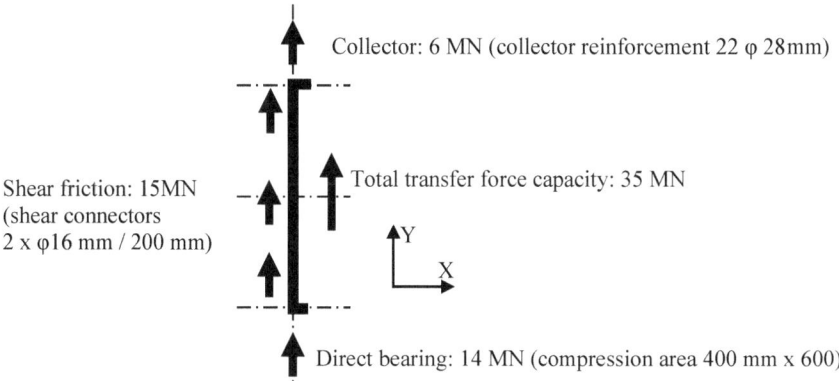

Fig. 5.33 Capacity of podium slab—shear wall connection

slab thickness of 400 mm, the following capacities can be computed for each mechanism:

- F_1 = 15 MN for shear friction, considering a slab connection reinforcement of $2 \times \varphi$ 16 mm/200 mm perpendicular to the shear wall. This capacity cannot be further increased by additional reinforcement because of the limitation of the concrete tensile strength. In order to increase the shear friction capacity, an increase in slab thickness is necessary.
- F_2 = 14 MN for direct bearing, considering a characteristic compressive strength of 40 MPa and a contact area between the slab and the web of the shear wall of 600 mm × 400 mm. This force can be increased just by increasing the quality of concrete or by increasing the thickness of the slab.
- F_3 = 6 MN through a collector reinforcement of 22 φ 28 mm rebars (Fig. 5.34).

The total transfer capacity is 35 MN. Because changing the concrete quality just for the podium slab is not a viable option in practical design and the podium slab is already rather thick, the total transfer capacity can be slightly increased just by a larger amount of collector reinforcement. However, detailing issues might occur

Fig. 5.34 Details for the collector reinforcement

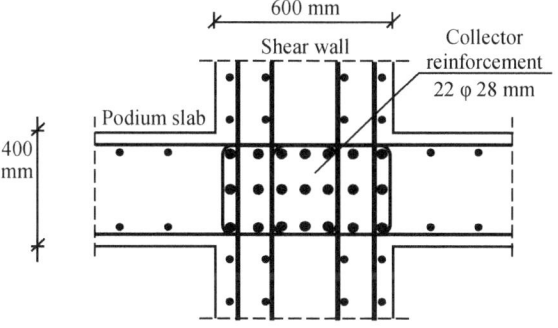

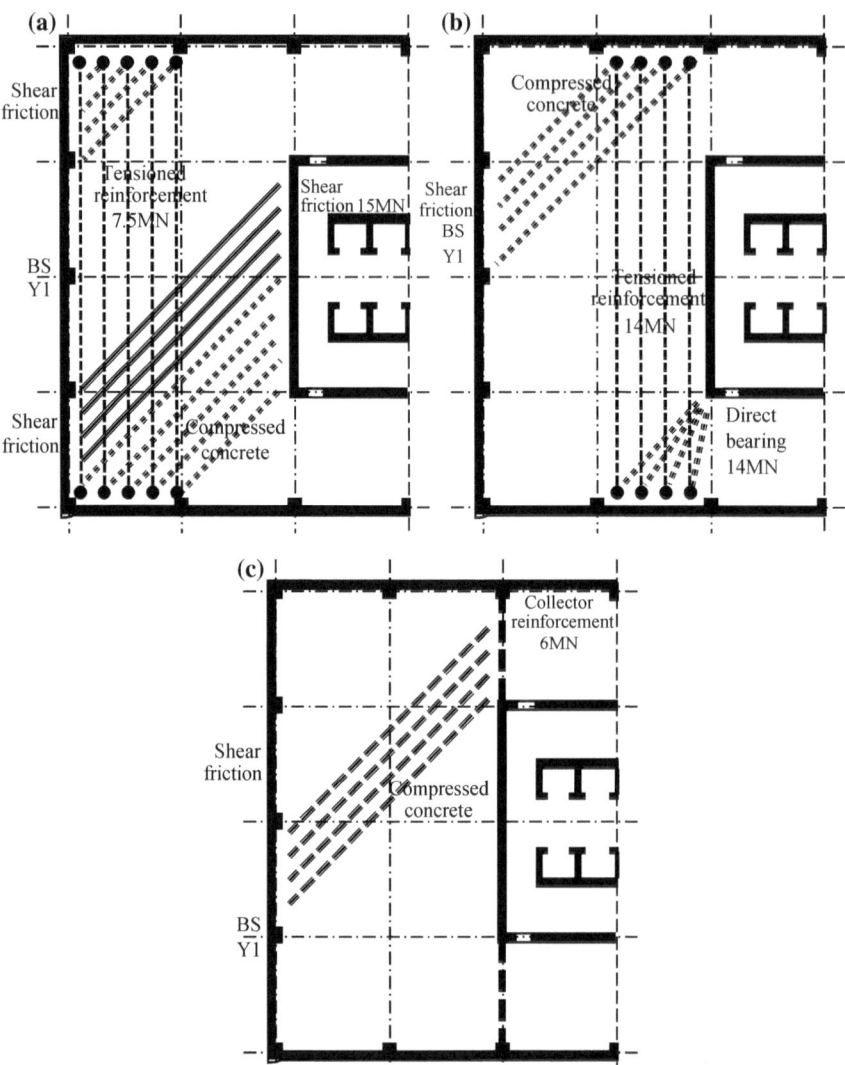

Fig. 5.35 Transfer of the shear force by shear friction (**a**), direct bearing (**b**) and collector reinforcement (**c**), through the podium slab, to the basement wall

when attempting to install more than 22 φ 28 mm rebars in the wall-slab intersection area.

The above computed forces need to be transferred through the slab to the outer basement wall aligned with the Y direction (labelled here BSY1). Load paths through the podium slab are schematically presented in Fig. 5.35. Half of $F_1 = 15$ MN is transferred directly by in-plane compression struts that find support exactly on the basement wall BSY1. The other half needs to be further suspended by slab reinforcement (Fig. 5.35a). The same amount of reinforcement as the connection reinforcement $2 \times \varphi$ 16 mm/200 mm is necessary to fully suspend 7.5 MN over 8.1 m length. To transfer the load from the compression struts in the slab to the basement wall BSY1 a $2 \times \varphi$ 16 mm/200 mm connection reinforcement is necessary.

Because of the slab geometry, direct bearing load F_2 cannot be directly transferred to the basement wall. From the wall-slab compression contact surface, this load is spread in the slab by in-plane compression struts and further transferred to the back of the slab by reinforcement ties aligned to the Y direction. Assuming a maximum inclination angle of the compression strut with respect to Y direction of 45°, the direct bearing force should be suspended using tie reinforcement distributed in the slab located roughly at a maximum 8.1 m from the wall's lateral face. This results in 2φ 16 mm spaced at 100 mm as presented in Fig. 5.35b.

Tensile force in the collector reinforcement, $F_3 = 6$ MN, can be directly transferred to the basement wall through the podium slab by in-plane diagonal compression struts (Fig. 5.35c).

It must be noted that any attempt to reduce the thickness of the slab in the vicinity of the central concrete core will result in a severe decrease in the slab capacity necessary to transfer the necessary shear forces. According to Romanian seismic design code P100-1/2013 (2013), a reduction of the slab thickness to 300 mm results in a decrease in the capacity to slab-wall connection from 35 to 27 MN.

This exercise demonstrates that it is necessary to use the full capacity of a thick podium slab so as to be able to transfer the shear forces at the podium level.

References

ACI 318-11 (2011) Building code requirements for structural concrete and commentary, Reported by ACI Committee 318

CEN (2004a) Eurocode 2: design of concrete structures—part 1-1: general rules and rules for buildings. European Standard EN 1992-1-1, Brussels

CEN (2004b) Eurocode 8: design of structures for earthquake resistance—part 1: general rules, seismic actions and rules for buildings. European Standard EN 1998-1, Brussels

Cotofana D, Pavel M, Popa V (2017) Anchorage of beam reinforcement in beam-column joints: overview of the Romanian standards. In: 6th national conference on earthquake engineering and 2nd national conference on earthquake engineering and seismology, Bucharest, Romania

Craciun I, Vacareanu R, Pavel F (2016) Spectral displacement demands for strong ground motions recorded during Vrancea intermediate-depth earthquakes. In: Vacareanu R, Ioenscu C (eds) The 1940 Vrancea earthquake. Issues, insights and lessons learnt. Springer natural hazards. pp 169–188. https://doi.org/10.1007/978-3-319-29844-3_12

D'Ayala D, Paganoni S (2009) Assessment and analysis of damage in L'Aquila historic city centre after 6th April 2009. Bull Earthq Eng 9(1):81–104. https://doi.org/10.1007/s10518-010-9224-4

Earthquake Engineering Research Institute—EERI (2012) Special earthquake report, The Mw 7.1 Erciş-Van, Turkey Earthquake of October 23, 2011

Lagomarsino S (2012) Damage assessment of churches after L'Aquila earthquake (2009). Bull Earthq Eng 10(1):73–92. https://doi.org/10.1007/s10518-011-9307-x

Lagos R (2017) The quest for resilience in seismic design of RC buildings: The Chilean practice. In: 16th world conference on earthquake engineering, Santiago de Chile, Chile, oral presentation

P100-1/2013 (2013) Code for seismic design—part I—design prescriptions for buildings. Ministry of Regional Development and Public Administration, Bucharest, Romania

Popa V (2014) Upgrade of P100-1 concrete provisions. Mathematical modeling in civil engineering, no 3/2014. https://doi.org/10.2478/mmce-2014-0014

Popa V, Văcăreanu R, Karadogan F (2013) Post-earthquake investigation and seismic evaluation of a damaged RC building in Van, Turkey. In: 10th international conference on urban earthquake engineering, Tokyo Institute of Technology, Tokyo, Japan, 1–2 Mar 2013

Popa V, Vacareanu R, Oprisoreanu V, Albota E, Kober D (2015) Suitability of current assessment techniques to retrodict the seismic damage of buildings. A case study in Van, Turkey. Open Civil Eng J 9:330–343. https://doi.org/10.2174/1874149501509010330

Vacareanu R, Marmureanu G, Pavel F, Neagu C, Cioflan CO, Aldea A (2014) Analysis of soil factor S using strong ground motions from Vrancea subcrustal seismic source. Rom Rep Phys 66:893–906

Chapter 6
Case Studies

Abstract In this section, an example of a life-cycle analysis and of a resilience analysis is given for a reinforced concrete high-rise building in Bucharest. The building was designed for three levels of peak ground acceleration in order to mimic the various seismic design code generations enforced in Romania. The economic feasibility of a further increase of the design peak ground acceleration is evaluated through life-cycle analysis. This type of analysis uses as input data the losses obtained from a seismic risk analysis performed with fragility functions specifically built for the three analysed structures. The results show that both from the economic point of view and from the functionality point of view it is feasible to further increase the design peak ground acceleration from the current hazard level that corresponds to a mean return period of 225 years.

Keywords Life-cycle analysis · Seismic resilience · Fragility functions Seismic risk · Nonlinear time history analysis · Intensity measure Damage state · Pushover curve

The seismic risk analysis of three 16-story symmetrical buildings with structural systems consisting of reinforced concrete (RC) shear walls is given in the subsequent section. The planar layout of the structures is shown in Fig. 6.1. The structural system consists of two 12 m RC structural walls placed in each of the two principal directions, columns and beams. The slab is made of RC and has a thickness of 15 cm. Concrete class C40/50 and reinforcing steel grade S500 are used for all the structural components. The height of all the stories is 3 m. The structural design was performed using the provisions of the current Romanian seismic design code P100-1/2013 (2013) for the seismic conditions of Bucharest (the values of the control periods T_C and T_D are 1, 6 and 2, 0 s, respectively). The structural behaviour factor q for RC structures with cantilever walls is equal to 4.0. The difference between the three structures is with regard to the value of the peak ground acceleration used for anchoring the normalized design acceleration response spectrum, that was considered as 0.24 g (value corresponding to the previous Romanian seismic design code), 0.30 g (value corresponding to the current Romanian seismic design code) and 0.36 g (value that has an exceedance

© The Author(s) 2018
F. Pavel et al., *Impact of Long-Period Ground Motions on Structural Design: A Case Study for Bucharest, Romania*, SpringerBriefs in Geotechnical and Earthquake Engineering, https://doi.org/10.1007/978-3-319-73402-6_6

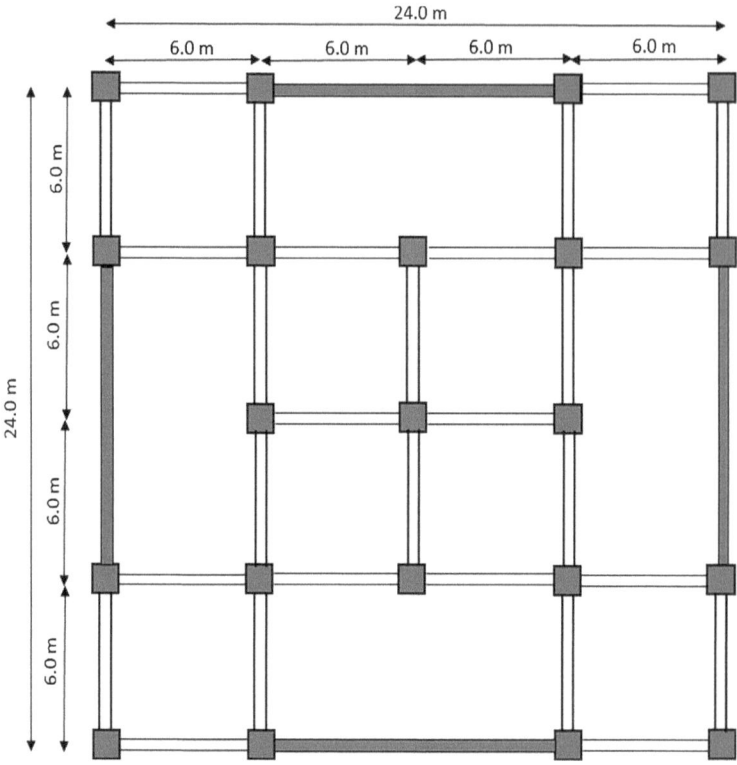

Fig. 6.1 Planar layout of the three analysed structures

probability of 10% in 50 years, according to the seismic hazard curve for Bucharest). In Table 6.1, the cross-sections of the main structural elements and the range of longitudinal reinforcement percentages for each component are given. The three structures will be denoted hereafter as S1 (structure designed for a design peak ground acceleration $PGA = 0.24$ g), S2 (structure designed for $PGA = 0.30$ g) and S3 (structure designed for $PGA = 0.36$ g). The fundamental vibration period for the three structures, using cracked RC sections for the modal analysis, is in the range 1.00–1.13 s.

The pushover analyses were performed using STERA3D software (http://www. rc.ace.tut.ac.jp/saito/software-e.html). The basic assumptions of the static nonlinear analyses (pushover) are described hereafter. All the structural elements are considered as line elements and the slabs are modelled as rigid in their plane. The beams are modelled as elements with nonlinear flexural hinges at both ends and a nonlinear shear hinge in the middle of the element. The column elements are modelled using a multi spring (MS) model with nonlinear axial springs at both ends and bi-directional nonlinear shear springs in the middle. The wall element is similar to the column element, consisting of both nonlinear axial springs and shear springs

Table 6.1 Characteristics of the structural components for the three analysed structures

Struct.	Structural element						
	Structural walls		Columns		Beams		
	Cross-section/ thickness	Long. reinforcement ratio (%)	Cross-section	Long. reinforcement ratio (%)	Cross-section	Long. reinforcement ratio (%)	
S1	Boundary— 60 × 60 cm Web—25 cm	Boundary—0.85% Web—0.27%	60 × 60 cm	0.85	30 × 60 cm	0.44–0.84	
S2	Boundary— 60 × 60 cm Web—30 cm	Boundary—0.85–1.69% Web—0.27–0.60%	60 × 60 cm	0.85	30 × 60 cm	0.56–1.10	
S3	Boundary— 60 × 60 cm Web—40 cm	Boundary—0.85–2.73% Web—0.25–0.77%	60 × 60 cm	0.85	30 × 60 cm	0.56–1.64	

in both the boundary elements and in the wall web. In addition, the wall element has rigid elements at the top and at the bottom part of the element for each story. The additional effect of the slab, as well as a fraction of its reinforcement, are taken into account in the modelling of the beam elements. The hysteretic model of the beam element is based on a trilinear model which can account for strength degradation, stiffness degradation and for slip degradation, as well. The multi spring (MS) model for the vertical elements consists of five areas—the four corner areas have both concrete and steel nonlinear springs, while the central area has only a nonlinear concrete spring. The concrete and steel strengths are based on mean strengths of the materials taking into account also the effects of the concrete confinement and the steel consolidation. The tension strength of the concrete is neglected in the analyses. In the case of the steel nonlinear springs, the maximum-oriented model is adopted prior to yielding and a trilinear model is adopted afterwards. The trilinear hysteresis rule is also adopted for the concrete springs with consideration of the strength degradation after the yielding point. A Rayleigh damping model is used in the computations, with a damping ratio of 5% for the first eigenmode.

In order to evaluate the seismic risk associated with the three 16-story buildings, we first obtained the fragility functions associated with four damage states, namely slight, moderate, extensive and complete. The drift limits for each of the four damage states were taken from HAZUS manual (2012) and correspond to a high-rise RC structural walls system. The fragility functions were obtained using the FRACAS approach, proposed by Rossetto et al. (2016). This approach makes use of inelastic response spectra of ground motion recordings in order to construct the fragility curves, while in the second approach the fragility functions are obtained through pushover analysis. An incremental dynamic analysis (IDA) curve is constructed by performing nonlinear time-history analyses of equivalent inelastic single-degree-of-freedom (*SDOF*) systems. The capacity curve of the three 16-story structures was approximated using an elastic-perfectly plastic (*EPP*) hysteretic model. The engineering demand parameter (*EDP*) under investigation is the maximum inter-story drift. Twenty horizontal components of ground motions recorded in the Bucharest area during the Vrancea earthquakes of March 1977 ($M_W = 7.4$, $h = 94$ km), August 1986 ($M_W = 7.1$, $h = 131$ km) and May 1990 ($M_W = 6.9$, $h = 91$ km) were considered as input for FRACAS. The 20 individual acceleration response spectra, as well as their mean and median spectral accelerations are plotted in Fig. 6.2.

The intensity measure (*IM*) of the recorded ground motions used in FRACAS is the peak ground acceleration, *PGA*. The parameters of the fragility functions are the median values of *PGA* expressed as fractions of g and the corresponding standard deviations. The values of the parameters of the fragility functions for each damage state are reported in Table 6.2. A comparison between the fragility functions computed for the four damage states is shown in Fig. 6.3.

The probabilities of being in certain damage states for various peak ground acceleration levels are given in Tables 6.3, 6.4 and 6.5. The probabilities associated with each damage state show that the overall structural performance improves, as expected, from structure S1 to structure S3. This fact is also confirmed by the sum

Fig. 6.2 Absolute
acceleration response spectra
for the 20 horizontal
components used for
nonlinear time-history
analyses of *SDOF* systems

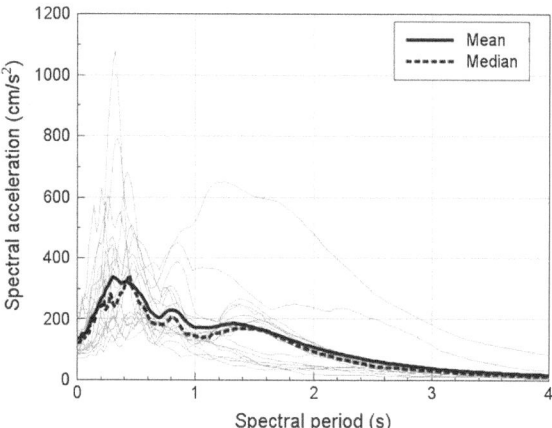

Table 6.2 Median and standard deviations of the fragility functions for each damage state
(FRACAS results)

Damage state	Structure					
	S1		S2		S3	
	Median ('g)	Standard deviation	Median ('g)	Standard deviation	Median ('g)	Standard deviation
Slight	0.077	0.665	0.086	0.762	0.108	0.695
Moderate	0.185	0.990	0.219	0.855	0.333	0.735
Extensive	0.568	0.523	0.607	0.546	0.625	0.619
Complete	0.774	0.381	0.855	0.391	0.978	0.491

of the probabilities associated with either none or slight damage state, that increases
again from structure S1 to structure S3 for all the analysed peak ground acceleration
levels.

The mean annual rate of exceeding a particular damage state is obtained by
convolving the seismic hazard curve with the fragility function corresponding to the
particular damage state. The mean annual rates of exceedance for each of the four
damage states are given in Table 6.6. Again, the same improvement in the overall
structural performance can be observed from structures S1 to S3.

Subsequently, the seismic performance for the three analysed structures was evaluated using a Monte Carlo simulated earthquake catalogue for the Vrancea
intermediate-depth seismic source. This catalogue was used previously by Pavel et al.
(2017) in the assessment of seismic risk for residential buildings in Bucharest. The
catalogue covers a period of 50,000 years and consists only of individual seismic
events with moment magnitudes $M_W \geq 6.0$. For each event, the probabilities of being
in a particular damage state were evaluated for each of the three analysed structures.
The HAZUS (2012) model for the estimation of losses and casualties was employed in
the computations. The mean normalized damage degree (on a scale from zero—no

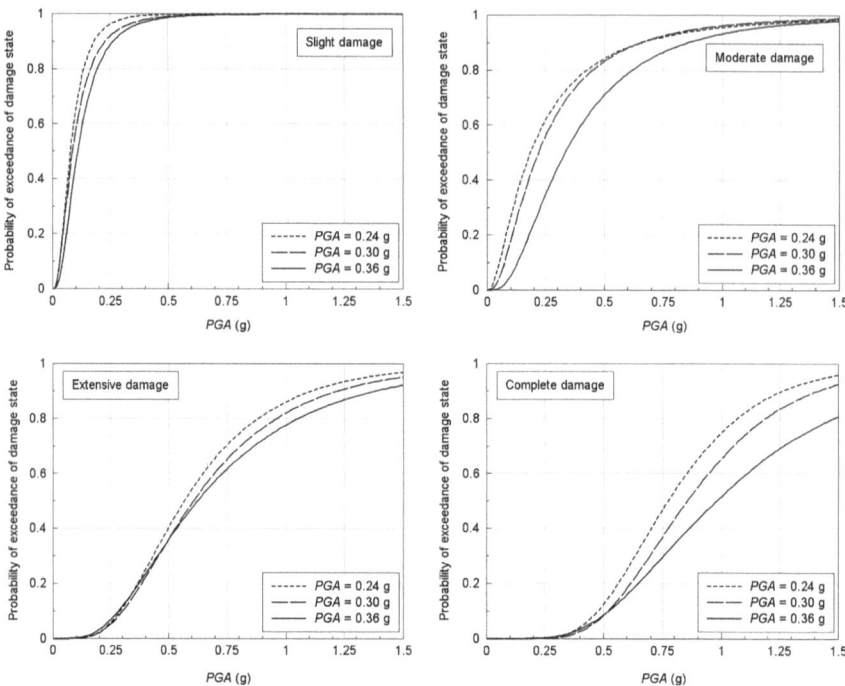

Fig. 6.3 Comparison of the fragility functions for the four damage states

Table 6.3 Probabilities (expressed in percentages) of being in certain damage states for various peak ground acceleration levels for structure S1

PGA	No damage (%)	Slight damage (%)	Moderate damage (%)	Extensive damage (%)	Complete damage (%)
0.10	34.71	38.57	26.67	0.04	0.00
0.15	15.80	42.59	41.07	0.54	0.00
0.20	7.56	39.30	50.84	2.28	0.02
0.25	3.83	34.22	56.12	5.68	0.15
0.30	2.04	29.22	57.62	10.47	0.64
0.35	1.14	24.84	56.29	15.87	1.86
0.40	0.66	21.14	53.07	20.97	4.16

damage—to one—collapse) for the entire earthquake catalogue is 0.167 for structure S1, 0.141 for structure S2 and 0.097 for structure S3.

The mean annual losses were subsequently used in the life-cycle analysis of the three structures (Kappos and Dimitrakopoulos 2008). The results of the life-cycle analysis are shown in Fig. 6.4. The discount rate was considered as 2, 4 and 6%, while the planning horizon has a duration between 0 and 50 years. The discount

Table 6.4 Probabilities of being in certain damage states for various peak ground acceleration levels for structure S2

PGA	No damage (%)	Slight damage (%)	Moderate damage (%)	Extensive damage (%)	Complete damage (%)
0.10	42.15	39.88	17.91	0.05	0.00
0.15	23.27	43.83	32.38	0.52	0.00
0.20	13.40	40.82	43.67	2.09	0.01
0.25	8.07	35.78	50.94	5.13	0.08
0.30	5.05	30.59	54.52	9.47	0.37
0.35	3.27	25.90	55.17	14.55	1.12
0.40	2.18	21.87	53.70	19.65	2.60

Table 6.5 Probabilities of being in certain damage states for various peak ground acceleration levels for structure S3

PGA	No damage (%)	Slight damage (%)	Moderate damage (%)	Extensive damage (%)	Complete damage (%)
0.10	54.41	40.51	4.93	0.15	0.00
0.15	31.82	54.28	12.84	1.05	0.01
0.20	18.76	56.84	21.11	3.22	0.06
0.25	11.36	53.82	27.89	6.67	0.27
0.30	7.08	48.57	32.57	10.98	0.80
0.35	4.53	42.77	35.25	15.63	1.82
0.40	2.98	37.17	36.30	20.11	3.43

Table 6.6 Mean annual rates of exceeding a particular damage state (FRACAS results)

Structure	Slight damage	Moderate damage	Extensive damage	Complete damage
S1	1.95E-01	8.12E-02	2.28E-03	4.64E-04
S2	1.85E-01	4.59E-02	2.06E-03	3.30E-04
S3	1.15E-01	1.47E-02	1.69E-03	1.98E-04

rates of 2 and 6% are more extreme cases, while the value of 4% is very close to the value used in Romania for infrastructure projects (4.4%). The results show very clearly that from the economic point of view, it is far more feasible to design the structure using an increased level of the seismic force for planning horizons in excess of 15 years (as it is normally the case for new buildings). In addition, it is also clear that an increase in the discount rate leads also to a decrease in the total losses during the same planning horizon.

Finally, in order to evaluate the recovery rate of the building functionality in the aftermath of a seismic event, a seismic resilience assessment was performed. The approach employed is the one proposed by Burton et al. (2015). In Fig. 6.5 the

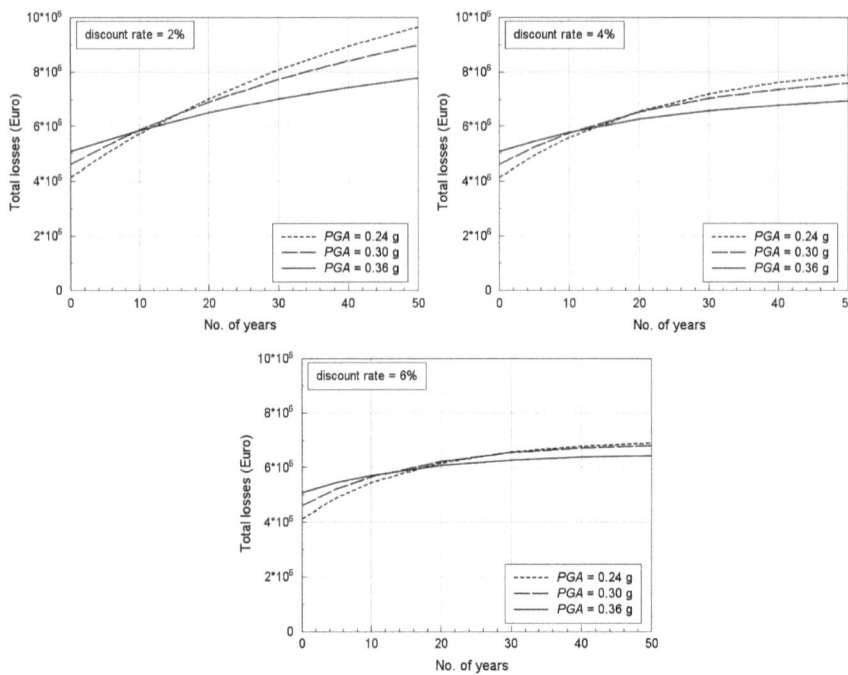

Fig. 6.4 Total losses for the three structures as a function of the discount rate and planning horizon

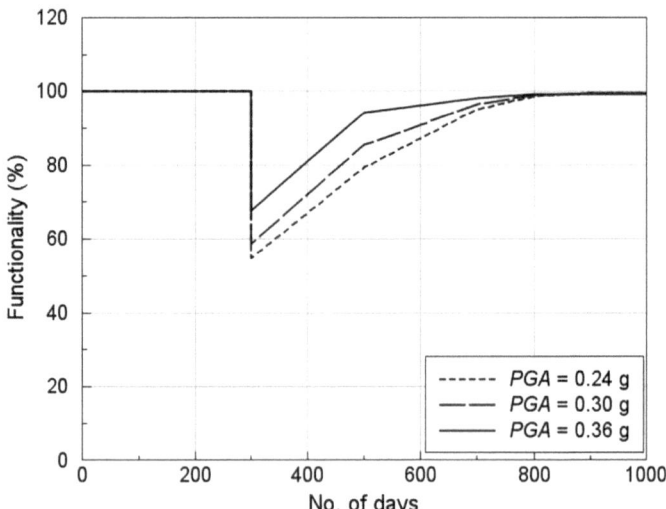

Fig. 6.5 Recovery curves for functionality for the three analysed structures

mean recovery curves obtained for the entire Monte Carlo simulated earthquake catalogue are compared. The time span in which 95% of the pre-earthquake building functionality was obtained varies from 402 days for structure S1 to 237 days for structure S3.

References

Burton HV, Deierlein G, Lallemant D, Lin T (2015) Framework for incorporating probabilistic building performance in the assessment of community seismic resilience. J Struct Eng 142: paper no. C4015007
Federal Emergency Management Agency (2012) Multi-hazard loss estimation methodology. Earthquake model—HAZUS MH 2.1. Technical Manual, Washington, USA
http://www.rc.ace.tut.ac.jp/saito/software-e.html
Kappos AJ, Dimitrakopoulos EG (2008) Feasibility of pre-earthquake strengthening of buildings based on cost-benefit and life-cycle cost analysis, with the aid of fragility curves. Nat Hazards 45:33–54
Pavel F, Vacareanu R, Calotescu I, Sandulescu AM, Arion C, Neagu C (2017) Impact of spatial correlation of ground motions on seismic damage for residential buildings in Bucharest, Romania. Nat Hazards 87:1167–1187
Rossetto T, Gehl P, Minas S, Galasso C, Duffour P, Douglas J, Cook O (2016) FRACAS: a capacity spectrum approach for seismic fragility assessment including record-to-record variability. Eng Struct 125:337–348

Chapter 7
Conclusions

Bucharest is among the European cities that have the highest seismic risk levels. The combination of its vulnerable building stock and its proximity to the Vrancea intermediate-depth seismic source are the two main factors responsible for the high level of seismic risk. The analysis of recorded ground motions during significant Vrancea intermediate-depth earthquakes in the past 40 years has shown that considerable long-period spectral amplifications may occur in certain situations. Unfortunately, the existing database of ground motion recordings from seismic events with $M_W \geq 7.0$ is extremely limited (only 13 recordings) and as such the evaluation of the causes of these long-period ground motions is an ongoing challenging research issue. Nevertheless, by combining the results of different studies from the literature, several issues related to the occurrence of long-period ground motions have been addressed.

The first main issue is related to the predominant periods of the soil layers beneath Bucharest. Several studies have pointed to the existence of significant spectral amplifications in the period range 1.2–1.6 s and at spectral periods in excess of 5.0 s. While the amplifications for the period range 1.2–1.6 s were confirmed by ground motion recordings, the amplifications for periods in excess of 5.0 s are visible only from the *HVSR* spectra.

The occurrence of long-period spectral ordinates is caused by a combination of large magnitude seismic events and relatively short source-to-site distances. This combination leads to a shift in the predominant periods of the acceleration response spectra towards the period range 0.8–2.0 s. It is true that large magnitude and more distant earthquakes can also generate considerable long-period spectral ordinates over a broad period range. In addition, another factor that has to be taken into account is the rupture propagation velocity, which can approach the shear-wave velocity limit, leading to an increase in seismic damage (as was the case of the Vrancea earthquake of March 1977). Another noteworthy aspect is related to the much larger spectral displacement demand observed from the ground motion recorded at INCERC station during the Vrancea 1977 seismic event, as compared to

© The Author(s) 2018
F. Pavel et al., *Impact of Long-Period Ground Motions on Structural Design: A Case Study for Bucharest, Romania*, SpringerBriefs in Geotechnical and Earthquake Engineering, https://doi.org/10.1007/978-3-319-73402-6_7

the spectral displacement values observed in Santiago (Chile) during the Maule (2010) earthquake. The same observation holds true if one compares the displacement demands from the same ground motion recording with the ones observed within the Wellington area during the Kaikoura (2016) earthquake. Moreover, one has to take into account the fact that the 1977 event was smaller in terms of magnitude as compared to the earthquakes of 1738, 1940 or 1802. Consequently, large displacement demands are to be expected within the Bucharest area in the event of a future large magnitude Vrancea intermediate-depth earthquake.

The soil response nonlinear analysis performed using the approach suggested by Pitilakis (2017) shows large long-period spectral ordinates that might even exceed the code spectral values. Moreover, the soil response nonlinear analysis confirms the significant long-period spectral amplifications which are to be expected during large magnitude intermediate-depth Vrancea earthquakes.

Due to the lack of available ground motion recordings from large magnitude Vrancea seismic events, the existing recordings were complemented with simulated ground motions that serve to fill in the gaps in the existing ground motion database. Moreover, a site-dependent ground motion model (applicable only for the Bucharest area) was derived from a database consisting of equal numbers of recorded and simulated ground motions. The resulting ground motion model can exhibit significant long-period spectral amplifications for large-magnitude earthquakes occurring at small source-to-site distances. In addition, because of the much better constrained database used in regression analysis, the variability of the proposed model is almost half that of the Vacareanu et al. (2015) model, which was used for the seismic hazard assessment at national scale.

A new probabilistic seismic hazard assessment was performed for Bucharest using the previously mentioned ground motion model. The analysis reveals much larger long-period spectral ordinates as compared to the ones obtained using the ground motion models applicable at regional scale, as performed in the study by Pavel et al. (2016). The hazard analysis shows that a larger variability can be expected in terms of spectral accelerations within Bucharest, as compared to spectral displacements.

The shape of the design acceleration spectrum in Bucharest has particular implications for the design and construction process. Seismic base shear coefficients of 0.2 are commonly used for the seismic design of mid-rise buildings in Bucharest. Such values are determined based on relatively large behaviour (reduction) factors ranging between 4 and 6. Capacity design procedures are strictly applied. Attempts to minimize the seismic damage to reinforced concrete buildings by reduction of the behaviour factors are not technically feasible for mid-rise buildings because of the large acceleration spectral ordinates. The large seismic design loads result in highly robust structures. Usually, in reinforced concrete shear wall structures the wall densities on the ground floor exceed 0.2%, calculated based on the total area of the building. The shear wall density is two times larger for RC structures built in Bucharest as compared to those built in Chile. Concrete and reinforcement consumption indexes in regular mid-rise buildings are significantly higher than those in other European countries that are not exposed to severe ground motions. Buildings need to be designed for large lateral displacement demands exceeding 60 cm. These large displacement demands raise serious concerns regarding the ductility of reinforced concrete elements, structural damage control and behaviour of nonstructural elements. Masonry partitions are still highly popular in Bucharest, particularly in residential buildings, increasing the potential seismic damage to new buildings in the case of large-magnitude Vrancea earthquakes.

A seismic risk analysis performed for three 16-story RC structural wall buildings, designed for the seismic soil conditions in Bucharest and for three levels of peak ground accelerations, namely 0.24, 0.30 and 0.36 g, has shown that long-term economic advantages might be gained if the buildings are designed for an increased level of peak ground acceleration. Thus, a further increase in the design peak ground acceleration from the current mean return period of 225 years to a mean return period of 475 years may be economically advantageous for planning horizons in excess of 15 years, irrespective of the value of the discount rate, increasing the seismic resilience of the building stock. Another advantage of increasing the design peak ground acceleration is related to a reduced degree of damage compared to lower design values, especially in the event of frequent seismic events.

As a general conclusion, the building stock of Bucharest is challenged by a unique combination of seismic source and site effects (at least in Europe), which generate very large displacement demands especially for mid-rise and high-rise buildings. These displacement demands give rise to many design and detailing issues for the structural engineers. The future revision of the Romanian seismic design code needs to address both the seismic hazard and the structural aspect in order to increase the level of safety and resilience for future buildings and their inhabitants.

References

Pavel F, Vacareanu R, Douglas J et al (2016) An updated probabilistic seismic hazard assessment for Romania and comparison with the approach and outcomes of the SHARE project. Pure Appl Geophys 173:1881–1905

Pitilakis K (2017) Site classification and definition of seismic actions in the revision of EC8. Keynote lecture presented during the 6th national Conference on Earthquake Engineering and 2nd national Conference on Earthquake Engineering and Seismology, Bucharest, Romania

Vacareanu R, Radulian M, Iancovici M, Pavel F, Neagu C (2015) Fore-arc and back-arc ground motion prediction model for Vrancea intermediate depth seismic source. J Earthq Eng 19:535–562